SNOW MEN

ANDREW CERONI

outskirtspress
DENVER, COLORADO

Snow Men

v4.0

Outskirts Press, Inc.
http://www.outskirtspress.com

ISBN: 978-1-4787-4457-3

PRINTED IN THE UNITED STATES OF AMERICA

For my sister, Nancy, forever and ever

Epigraph

Survivors aren't always the strongest; sometimes they're the smartest, but more often simply the luckiest.

Carrie Ryan
The Dark and Hollow Places

PROLOGUE

Eastern Chugach Mountains, Alaska
November

His hands still trembled. He couldn't shake it. He tried to spit and found he couldn't; he had no saliva. His tongue was a dry lump in a parched mouth. He poked the fire with a stick, stirring the embers to coax out a little comfort. Hell, he'd never been spooked like this before, never in his life.

Jack Stroud squatted in the snow, his rifle across his lap. He reached into his parka and pulled out the bottle. Nearly empty. Well, he still had another. A long, easy swig of Canadian whiskey felt oh so good going down, warmed his gut. He tossed the empty bottle in the snow.

Jack pulled a fresh bottle from his other pocket. Another swig. Nice. Jack knew he drank a lot. So what, it eased his pain. He really couldn't give a shit what other people thought. For more than six months, he'd watched his wife wither away from cancer. It was inoperable, nothing all those doctors could do. He loved her dearly, so much so the ache had never left him. Maybe her death was penance for his innumerable sins. He was never the best husband. All he had left for solace now were the wilderness and the whiskey.

He glanced skyward through the skinny arms of trees. The light was leaving the forest, and the winter night closing in.

Moon would rise soon, and it would be almost full. Even so, he knew he'd never get back to his base camp tonight. Better to stay here near the fire until daybreak, the rifle close by. No critters would come near the fire. It'd be down to near zero in no time. He zipped up his parka, and huddled closer to the fire.

He'd lost a glove too, dammit. It disappeared in the snow when he flung it down against the tree to take the shot. His hand was still oozing blood where he'd scraped the bark. The whole thing had been so… surreal. They'd surrounded him so suddenly, running upright too but… crazy. What the hell, he'd never seen anything like it. The whisky had nothing to do with it. He saw what he saw, whatever the hell it was. But he was pretty sure he hit one of them. He saw it drop. That made them back off. They melted into the woods in a hurry after he took the shot.

Soon after that he heard grunts, guttural sounds. Maybe some kind of beast talk. So perhaps he hurt it pretty bad. They were big creatures, not much shape to them, all white except for their huge, shiny black eyes. For their size, they moved very fast, tree to tree. There were several of them, but he didn't have wits enough to count. There were no polar bears in the Wrangell-St. Elias wilderness, and polar bears didn't run in packs anyway. And they didn't run on two legs either. So, what were they?

Jack had known something was off, not right, since yesterday afternoon. The forest seemed empty. He'd seen a couple of grizzlies out here several times during the last couple years, two bears that never slept through the whole winter. They were almost like brothers, never too far from each other. He'd watched them foraging, though most often the bears gave Jack a wide berth. And he knew of a good-sized wolf pack that had claimed the area northeast of Valdez. But now he'd seen no sign of anything; no bear sign, no wolf tracks. Nothing. The forest had cleared out, even the birds. Most birds went south in winter, but even in the coldest month, Jack very often saw ravens, waxwings, pine grosbeaks, even gyrfalcons flying around. Okay, he'd seen a few the last couple of days, but not nearly in their usual numbers. Well, perhaps all the wild things got spooked like he did.

His brother-in-law's words rang in his ears. *Jack, you're crazy to go out in the wilderness alone in the winter. Suppose you twist an ankle, break a leg, get attacked by wolves. Gees, how the hell is anybody gonna find you? We worry about you, that's all.*

Ed was a good guy, just a pain in the ass occasionally. On way too many occasions, Ed tried to give Jack advice he neither needed nor wanted. Ed ran an outfitter store in town, so Jack had to admit that Ed knew what he was talking about. It could be dangerous, being alone in the Chugach. But right now Jack wouldn't mind having Ed sitting across the fire with another rifle. Ed or, for that matter, anybody else would be a comfort now. Suppose those things come back? Jack shivered inside his parka. He only had so much ammo.

Winter nights in the Alaskan wilderness could be strange. Frigid, subzero air, an icy mist hovering over the ground. So cold that sap still lingering in a tree branch froze and snapped the branch, making a crack that reverberated in the dead silence of the woods. If you didn't know what it was, that could keep you awake all night. Then there was the dim light from a sliver of moon stretching long shadows across the snow. You could see things, things that weren't there. Even worse was wondering whether they're there or not. In the dead of night, the winter wilderness could spook even the well initiated.

Well, he wasn't hallucinating about those things today. They were there. No doubt about it. It wasn't the whiskey. He tossed another log on the fire. He would clear out of here as soon as the gray light of morning gathered on the horizon. Get back to the road and his truck. Go see the sheriff, first thing. He glanced at the rifle and pulled back the bolt to ensure a round was chambered. It was. He put the gloveless hand in his pocket.

Hours ran into each other. Jack dozed in and out of awareness. The fire was going down, but the parka kept him warm enough. Then… something… he jerked awake. What? He forced his eyes open, and slowly raised his left arm to see the watch. 3:00 AM. There it was again. The sound was unmistakable, more subtle than a tree branch popping. A footfall. A crunch of snow, somewhere to his left. Not far. Jack squinted, trying to see beyond the dim glow of the fire.

He slowly moved his right hand from his pocket to the rifle. He snapped off the safety, his finger curving around the trigger. He had one round in the breech, three in the magazine. If it was those damn creatures again, reloading was out of the question. He'd seen them run. They were fast. He was a crack shot, but it'd be tough to put the rifle's sights on one in this dark. After he expended the four rounds, he'd have to beat feet, and fast. West, if he could figure out where west was in the darkness.

Another crunch of snow. Shit, this one was to his right. They were surrounding him. Little by little, his right boot pushed a growing mound of snow toward the fire. It began to sizzle, steam floating into the air as the light diminished. His boot continued pushing the snow through the center of the dimming coals. He caught a movement straight ahead between two black spruce. Twenty yards away, if that. He rubbed his eyes to free them from the haze of sleep and whiskey. The creature was all white except for the large, glassy black eyes.

Now! Jack leaped to his feet, swinging the butt of the rifle to his shoulder. The beast leaped sideways to a tree. The sights of Jack's rifle followed the movement. A sudden blast of sound and light filled the forest, shattering the stillness and reflecting off the nearby trees. He swung the barrel immediately to his right. Another blast and flash of light from the muzzle. Then, two last blasts to his left. He threw the rifle with all the strength he could muster at what he thought was moving to his right and he heard a loud satisfying grunt of pain. There were more guttural sounds he couldn't decipher, that weird beast speak. Jack reached down for his backpack, slung it over a shoulder, and spun to his rear in one motion. West. He took off through the snow, struggling with the depth. The snowshoes were back by the fire.

As his eyes adjusted, he could somewhat make out a path in the moonlight. He reached under his parka for the hunting knife. Behind him, he now heard what sounded like shouts, not grunts. What the hell? He remembered a slight ravine, an old stream bed coming up on his left. He turned down it, changing direction to the south, toward the bay. The ravine sheltered his image as he dropped below the level of the woods. The sounds of pursuit, pounding footfalls, faded. Jack stopped abruptly and crouched in the low scrub, his heart pounding. Dead silence. A bead of sweat dripped into his eye. He rose gradually and tiptoed forward, desperate to put more distance between himself and his pursuers.

Faintly, he heard them again… guttural sounds, gibberish, growing in intensity. He gathered his strength and leapt forward again, his feet pounding through the snow and scrub, his lungs burning. The end of the ravine opened up ahead. As he passed out of the ditch, he spun right, back again west. But, there was no path here. Rough going. Jack stumbled between two trees. A mistake.

"Ahh!" he shouted as the wire from his own snare tightened around his left foot. "Damn!" He went down hard, but quickly jammed his knife between boot and wire until it snapped. Up again, slipping and scrambling in the snow. To the right, he saw two large white figures crashing through the woods, running parallel to him. He could hear more of them just behind. He raced on.

Just ahead he saw a break in the black spruce. Well, this would have to be it. He'd have to make a stand. He couldn't outrun them, not all night. As he reached the small clearing, he spun, his knife glinting in the pale light. He swung it back and forth, menacing. The creatures materialized between the trees from all directions.

"Come on, you assholes! Come to Jack! Come get it!" he screamed, the shriek bouncing off the crusts of snow.

As they drew in closer, looming over him, his arms were suddenly jerked behind him. He could feel a bone in his thumb snap as the knife was struck from his hand. The pack was wrenched from his shoulder and tossed on the ground. Without warning, he was hit hard in the gut. Again, and then a third time. Jack fell to his knees in the snow, nauseous and gasping for air. He wretched, then twisted, trying to see his attacker, and stared into the strange, glassy black eyes. Goggles!

"Who the hell are you? What do you want?" Jack shouted, baring his teeth. He writhed in the iron grip and glared around the circle.

A white-gloved hand drew close to his face as the figure loomed over him. A bright titanium blade suddenly thrust forward and entered Jack's neck under his right jaw. It ripped forward and left, slicing through the right branch of the carotid artery. Jack's eyes opened wide in horror. He tried to speak but couldn't as air rushed out from his lungs and torrents of warm fluid coursed down his neck. Why? Jack slowly shook his head right, then left. His vision blurred and he felt himself sinking downward, his face settling into the snow and his own blood. Then blackness. His boots were toe down, sticking precariously up and out of the snow.

The black-goggled faces glanced at one another, then backed away from Jack Stroud's body. His life's blood pooled about his head, black in the pale moonlight. The same hand that had cut Jack's throat rose in the night air and waved, pointing eastward. One of the white figures, his chin bruised and bloodied from being hit by Jack's thrown rifle, dropped the rifle in the snow next to his body. They drew away, their boots crunching in the snow, and disappeared east into the woods.

CHAPTER 1

Colorado Springs, Colorado
November

The Toyota car door edged open and Pete Novak's cordovan dress boot reached for the concrete drive. His shoulder holster thumped against his chest as he climbed out of the rental car. He scrunched his eyes, his forehead wrinkling. Novak hoped his sudden appearance wouldn't spook McClure. It had been well over ten years since they'd last seen one another, so Pete was unsure what McClure's reaction would be. But, he had to give it a try. It was too urgent not to, the operation was too pressing. The Agency needed McClure's skill sets. And with the action likely to take place inside the United States, leadership at Langley Center knew it would require a delicate touch. Dave McClure could do it.

It had taken Novak forty-five minutes to drive from the Colorado Springs airport and finally find the place just up the hill and on the right on Allegheny Drive. Pete Novak stood in the driveway and assessed the house. Not bad. A two-story tan brick with a Spanish tile roof, pale burgundy trim, new thermal windows and a two-car garage. A good-sized stand of aspen stood just west of the house. Pete could imagine their bright gold leaves fluttering in an early autumn breeze, but now their bare branches raked the blue sky.

He glanced next door where an inflated Santa and snow man bobbed gently on the front lawn. *Seems people decorate for the holidays earlier and earlier each year,* he mused. Novak smiled when he noticed the exterior storm door was made of bronzed, welded steel with a matching steel frame. Two locks on the storm door, one a deadbolt. He could also see two locks, one of them also a deadbolt, on the interior wooden door. McClure was still security conscious.

The soles of his boots scraped on the cement as he sauntered up the three steps to the door. With a cool November wind rustling his sandy hair, tufts of gray showing on the sides, Pete reached out and pressed the doorbell. A few seconds and the interior door creaked open.

"Hi, how can I help you?" A look of anticipation washed across Dave McClure's face.

"Dave, I'm Pete Novak. Remember me?"

McClure's face brightened. "Mr. Novak! Wow, what a surprise!"

"So, you do remember. That's good. Dave, please, call me Pete."

"That'll be a struggle, but I'll try," Dave laughed. As the Deputy Director for Operations, or DepOps as he was commonly known at Langley Center, Pete Novak engendered respect more than familiarity. "Please, come in." McClure reached down and to the side for the key, unlocked the steel security door, and stepped aside.

"What brings you to Colorado Springs, Mr. N..., ah, Pete?"

After Pete stepped by him, McClure locked the doors and motioned to the family room around a corner.

"Coffee, Pete?"

"No, too late in the day. Thanks, though."

"Well, how about a 17-year-old scotch on the rocks?" McClure smiled as if he knew the answer.

"Hey, now that sounds inviting!" Novak laughed.

Scotches in hand, they sat down on opposite sides of the coffee table. Even in a baggy sweatshirt and jeans, McClure's trim fitness was evident. In his mid-thirties, McClure's curly brown hair and crystal blue eyes had ten years ago drawn the attention every young woman in the field operations school. Novak could see he hadn't changed much since then. Pete himself felt a little self-conscious about the little paunch he'd developed over the years. Then again, he wasn't in his thirties anymore either. He noticed the TV was on, but muted.

"What brings you to Colorado Springs, Pete? Business?"

"Oh yes, a meeting at Air Force Space Command tomorrow morning. I just got in, so I thought I'd look you up first. The research staff did some checking, told me where I'd find you. It's been over ten years, Dave. And by the way, you look in great shape."

"I still work out every day. Jog first thing in the morning, then the Bowflex. And yes, it's been a long time for sure. I'm sure you know I left the Bureau."

"I do. That's partly why I'm here. I heard about your wife and son. It's been a year now, hasn't it? Dave, I can't tell you how sorry I am about that. I'm truly sorry for you. I was told it was a very bad accident in the mountains."

A shudder went through McClure, his eyes grew distant. "Yeah, that's right. I was away in Cleveland running leads on a domestic terrorism investigation." He cleared his throat. "Anne and Michael… they were driving back from a soccer game in Grand Junction, on I-70 between Eagle and Vail. It was a micro-blizzard out of nowhere, getting dark, snowing, and the highway patrol said the roads were icing. Well, this 18-wheeler headed west was going way too fast to make the curve. It couldn't hold the road, blew through the median and slammed into the 4Runner." His eyes were wet. "They didn't have a chance..." He coughed and stared into his glass.

"I am so sorry, Dave."

"Asshole truck driver. He's dead too. So, as much as I want to, I can't put a bullet in his head for killing my family. I'm sorry Pete, it's still very much with me. A year now and it's still fresh in me."

"Dave, nobody gets over something like that easily. I expect that to some degree, it will always be with you. You have to remember the good times too. You have to move on."

McClure was near tears. "How the hell can I do that? Every time I walk down a hallway here... or enter Michael's room, I see their faces. I hear their voices. How can I ever get away from that?"

"You don't get away from it, you learn to deal with it. The way to do that will come to you in time. Look, so, you left the Bureau. What now?"

"I don't know. I can't do that anymore. I keep thinking that if I hadn't been away in..."

Novak interrupted. "No. Don't. Lose that thought now, Dave. That wouldn't have changed anything. It... it was their time. Let that go now."

"That what you really believe?"

"Completely." Novak took a sip of scotch. "Dave, I can't put myself in your shoes; that's impossible. I can only begin to imagine how gut-wrenching this last year has been for you. But, you're alive and your family would want you to move on in life."

"I understand that. I just don't know how to do it."

"That's partly why I'm here. Like you said, it's been a year. You need to think about going back to work. Maybe you can't see it now, but I'm sure it would help you. You need to climb out of this abyss."

"Don't think I haven't thought of it, but..."

"But what? You were one of the brightest students we'd ever seen in the Field Operations School. J.T. Brannon insisted that you be assigned to him in special operations. J.T. was sold on you! He said you were the best. He claimed you were a genuine J.T. Brannon, Jr.!"

"J.T. Brannon? Good God, I really liked that guy. He said that about me?"

"As J.T. would say, 'absofreakinglutely!'"

"Gees, Brannon. Where is J.T. now?"

Novak's eyes lowered a bit. "J.T.'s dead. A special mission that ended in a firefight. He took four bad guys with him. He was buried with honors in Arlington, and everyone in the 'business' attended. Fitting farewell for a genuine warrior. Like you, Dave."

McClure's brow furrowed. "Gee, Pete, I'm sorry to hear that. Brannon was a legend. All the instructors told us stories about him. I don't know, me and J.T., you're putting me in awfully rarified company."

"Come back to work for us. I'll see to it that you're immediately assigned to the field operations. We call it 'the Clandestine Service' now. I didn't just come to visit and check up on you, but also see if you'd be interested in coming back. We have a special need right now, urgent really, for someone who isn't officially a CIA asset, but has the skill sets of one. You're really the only guy I could think of. What do you think?"

"Where's the job."

"In the states. Southwest. I really can't be more specific than that. But the need… like I said, it's immediate. Interested?"

"Here in the states? That's unusual for the Company."

"Yeah, I know. But it's legit. A delicate situation. We're really the only ones who can handle it discreetly. Well, whaddya say?"

Pete Novak, the DepOps himself, was making him an offer.

Dave bet that didn't happen often. But he knew he couldn't accept, not right now. "Ah, I can't. Thanks for the offer, but tomorrow evening I'm leaving for Alaska. Bags are in the bedroom and already packed. Sorry."

Looking into Dave's eyes, Novak downed the last of the scotch, considering. McClure had blue eyes, extraordinarily blue. Novak remembered meeting Anne years ago. Jokingly, she had related to him that those crystal blue eyes were what had captured her initial attention in Dave.

"Alaska? Now? It's early November. 'Probably already snow all over the place. Pardon me for asking, but why Alaska?"

"It's my idea to maybe... well, to do what you said I should. Find some way to put all this... the pain and memory behind me." McClure smiled, "Alaska was one of the most enjoyable vacations we ever had together, Anne and I. Fairbanks, Denali, Anchorage, Seward and Valdez... forests, rivers, mountains, glaciers. Beyond beautiful. We enjoyed it so much. I'm hoping that going back will help me relive the good times. Find a way to mend things in my mind. Move on, you know, like you said I should."

Pete smiled. "Well, it's early winter up there now. What're you going to do?"

"Fly into Anchorage. Drive to Valdez. I've got a good hunting rifle, .300 Weatherby magnum, a .44 magnum handgun, knife, a pair of snowshoes. I'll get some ammo and provisions there, then hike into the forest northeast of Valdez. Provisions will last for a few days and after that I'll live off the land. It's the only thing I can think of that's so radically different from this place to get my mind off these things. I know Anne would want me to."

"Okay, but what if it doesn't work, Dave? What if it doesn't mend things? What if you find you can't live off the land? Then what? Are you sure you've thought this through? Coming back to work at Langley might do the same for you, even more so. Get your mind off things."

"Again, thanks for the offer, Pete, but I need to do this. I've already got the airline tickets."

"Look, please answer my question. What if you can't shake this while you're up there knee deep in snow? More importantly, can't live off the land?"

Dave's face grew solemn. "Pete, my mind's made up. And to answer your question, there are worse places to die."

"I see. I was afraid of that." Novak's face grew stern, discouraged to hear McClure might be willing to make this a one-way trip.

"Don't be. Really, I have to do this."

Novak stood, resigned but smiling. "Well, Dave, at least promise that you'll call me when you get back. You'll get back all right. You're smart. Tough too. Yeah, another Brannon."

They shook hands over the coffee table, and Dave nodded. "Count on it, Pete. And thanks for understanding." He pointed to the TV. "You see the news yet today? Iran's announcement that she's opening all her nuclear sites, the whole country, to UN inspectors? That she's decided to renounce her quest for development of weapons grade nuclear material? Sounds like it caught the whole damn world by surprise. Israel has already stated she's not buying it though. What do you make of it"

"I saw it on the plane coming out. Yes, it caught everybody off guard. We don't know what to think, but I'm fairly sure I don't buy it myself. The meeting at Air Force Space Command may be abbreviated. The Watch Center called while I was on my way up here to see you and said the Director wants me back pronto. So, we'll see."

"Good luck."

"Yeah, no kidding. I have a feeling we may need it."

They walked to the front door and shook hands again.

Novak turned, his eyes drilling into McClure's. "Dave, you watch yourself up there in the wilderness. It'll be an icebox this time of year. Be damn careful. And don't be fatalistic about the future. You have friends here. Remember that. Call me when you get back. I've got work for you." he paused. "Oh, some last words of advice from an old big game hunter, my father... he bagged a few moose up there. If anything does go wrong up there, you know, like it all suddenly becomes a survival issue, get to the coast if you can and light a big fire. The coast is always the best place to be seen by a ship or plane."

"Will do." McClure nodded again. "Thanks for coming by, and thanks for the words too, Pete. Really."

McClure watched until Novak's rental car eased down the driveway and backed into the street. *Interesting,* he thought to himself. *The CIA...Pete Novak. They searched for me, found me and Novak himself came all the way out here offer me a position. Interesting.*

CHAPTER 2

Six Weeks Earlier, September
Rive Gauche, Paris

The wet sheen of an early morning drizzle had dried on the streets hours ago. The sun was out and the air had pleasantly warmed up on the *Rive Gauche* in Paris. Kids were back to school all over Europe, but the young tourists and their older retired counterparts still loved to walk the streets on the Left Bank. Couples chatted as they strolled through Monmartre up to the *Sacré-Cœur Basilica* at the peak of the bluff. Later, they often lingered to browse the paintings, sketches and lithographs on display in the *Place des Artistes*. The treed plaza was ringed with bistros, where visitors could sip coffee or Chablis and just watch people for hours.

One of the most popular places for lunch was the nearby *Café Procope*, world renowned for its location and cuisine, but it's biggest claim to fame was for having the original writing table of Voltaire on display on the second floor. The classy black-lacquer exterior with the name in raised gold lettering exuded a sense of class and sophistication even before customers got to the door. Inside was no disappointment either: white tablecloths, pristine glassware, and red leather chairs, finished off with gorgeous 18th and 19th century art adorning the walls under glittering crystal chandeliers. Simply stated, a classy place.

The café was crowded, noisy at lunchtime. The multitude of tourists were so distracted by the wine, fine food and busy clatter of the place, they would later remember nothing of other patrons' faces, nothing out of the ordinary. This made the Café Procope perfect for a meeting of a discreet nature.

A black BMW 750li whispered through the right turn from *Rue de Buci* onto the *Rue de l'Ancienne Comédie*, its reflection fleeting in the storefront windows. Despite its sleek appearance, the car was heavily armored, sporting bulletproof glass and run-flat tires. The vehicle veered to the left as it approached the Café Procope and purred to a gentle halt. The right-front-seat shooter exited first, watchfully scanning both sides of the street for several minutes. Satisfied the area was secure, he opened the door and a middle-aged, elegantly dressed gentleman stepped gingerly out onto the street. His coal-black hair was slicked down, and the mustache, short beard and thick black eyebrows were impeccably manicured. A black cashmere topcoat, white shirt and black necktie completed the look for Bizhan Madani, chief of special operations for *Vevak*, the Iranian Intelligence Service. Immediately, a second shooter exited the right rear seat and approached the Café Procope. The BMW drew away and motored up the boulevard.

Following his armed escort, Madani stepped into the café. Eyes sweeping the room, the bodyguard nodded to Madani, then again in the direction of a corner table in the rear. Leonide Krasnoff nodded back. The shooter moved aside as Madani smiled and brushed past him toward Krasnoff's table. From his vantage point just inside the door, his keen gaze locked on two broad-shouldered men just one table away from the entrance. Craggy features and obvious bulges under their jackets gave them away: escorts for Krasnoff, Russian mafia. They caught each other's eyes, gesturing in acknowledgement.

"*Bonjour, mon ami. Ça va?*" Leonide rose to greet Madani, his hand extended.

"Oui, merci. Ça va bien." They shook hands and Madani settled into his seat. Leonide slipped off his gray wool coat as he sat. Only in his late 50s, Leonide's hair was already bone white, as was his bushy mustache. Broad shouldered and rugged looking, his face was furrowed from fifteen years as a Russian Mafioso, each year wondering if he would live to see the next. A glass of pale Chablis was nestled in front of him.

A waiter approached Madani. "Monsieur?"

"Oh, please, a cup of hot tea and a croissant."

"Lemon, monsieur?"

"No thank you."

"Oui." The waiter turned to the bar.

Krasnoff focused on Madani. "Bizhan, my friend, thank you for coming. I'm hoping we can come to closure on such an important proposition. So, Tehran is favorable to our offer, yes?" His warm welcome and smooth voice masked a deep distrust of Bizhan and all his Vevak kin. They were known for bloodshed, selling out and reneging on deals. Real dirt bags. But Krasnoff also knew that properly managed and with the right security forces as back-up, the deal at hand could be extremely lucrative for both him and his organization.

"Leonide, if such a thing is actually possible, then of course we are. But how such a thing may be achieved has us curious. So, my instructions are to ask you, how do you propose to accomplish this under the noses of the Americans?" Madani paused, waiting for an answer, then continued casually before Krasnoff could speak. "How? By the way, we already know your brother, Vassily, is a major in the GRU, now in SVR, and is working special operations with Spetsnaz. So, is Vassily involved in your plan? We would be much more confident if we knew the SVR was cooperating."

Krasnoff grimaced. "Bizhan, this we cannot discuss. Please…

do not mention my brother Vassily again. He has no connection with this whatsoever. The SVR is a bunch of paranoid bastards... that feverish, paranoid anxiety that was passed on to them by their KGB fathers. Vassily must constantly watch his step, as do all in the SVR. I must ask your cooperation in this. He needs to remain untainted in this matter." Krasnoff's eyes glittered while he spoke, but he smiled inwardly because in fact, Major Vassily Krasnoff was instrumental to the Russian plan. But this was not something Madani needed to know.

"Okay. I won't. I promise." Madani sighed and raised his hands in mock defeat. His tea delivered, Madani took a sip and munched on the croissant. Dabbing delicately at his mouth, he probed Leonide's eyes. The truth was, Madani trusted Krasnoff and his Russian Mafioso brethren even less than Krasnoff did him. Vevak's assessment that Russian gangsters were nothing but gutter dwellers was shared by many. This tenuous partnership was one of simple necessity. Madani took another sip of tea.

"Bizhan, thank you for understanding. Vassily must not be involved." Leonide gestured broadly with his right palm. "And, if I told you 'how', then you wouldn't need us, would you? Oh, you'd go and try to do it yourself, although your effort would be unsuccessful. So, no, I can't do that. But, the fact is, we can deliver what we have promised. So, the remaining questions are, is the price of 200 million U.S. dollars acceptable to Tehran? And, just as important, can Iran successfully deploy the warhead when we deliver it? I know that's not part of the deal, but I also have been asked to confirm these things."

"200 million is agreed as acceptable, Leonide. Ah, you said this will be a MIRV warhead, approximately 57 kilograms, yes?"

Despite the noise level in the café, Krasnoff lowered his voice, leaning across the table. "Okay, the price then is good, 25 million up front, the remainder upon delivery. Do you have the deposit?"

"Of course. Kasim is by the door. He will hand your people the briefcase upon your departure. And, the MIRV…?"

"My people…? You are observant."

"I have to be in my business. Kasim preceded me here by at least an hour."

"I see. Well, it's the same in my business. And now, Bizhan, the technical aspects. Yes, the warhead will be a multiple independently targetable reentry vehicle, a MIRV. Three reentry vehicles within the single warhead of about 450 kilotons each. Hiroshima was only about 16 kilotons, so this is a magnificent weapon, three times 450 kilotons! You will be able to target three cities with the one launch. And yes, the Minuteman III warhead itself is only 56.7 kilograms."

Some ten cars down from the Café Procope and on opposite side of the street, a silver Audi nudged into the curb. Two men were in front. The driver, Raphael, pressed a key on his cell phone. "Dani, you there?"

"Yes, at the bar, with Simone all over me. She's kissing my neck, nibbling my ear."

"She's supposed to. Besides, we all know you love it. Is Krasnoff still there?"

"Yes. The Iranian is at his table. They act like they know each other. But Krasnoff is doing most of the talking."

"Ah, did you get photos?"

"Yes, two. A frontal of Krasnoff, and an oblique profile of Madani. They're good shots."

"Not noticed?"

"No, very clean. By the way, there are more spooks in this place than I can count, even not including us."

"Well, Paris has become the new Berlin. Just the way the game is now. Except for us, Dani, this is no game. Any audio?"

"No, too noisy. I'd be noticed if I tried to get closer. And, I can't get a good angle on Madani's lips."

"Okay. "How about on Krasnoff? Everyone says your lip reading is the best... pick up anything?"

"Yes, some numbers, several of them, but I couldn't make it all out... 25, maybe 450." And, 'deposit,' that was clear. Also, maybe the name 'Marv' or 'Merv,' I'm not sure. So, there may be another major player involved."

"Very good work, Dani. Just a few minutes longer." But Raphael's brow wrinkled as he grew even more uneasy.

Madani's cheeks creased in a broad smile. "Hah! Leonide, this is fantastic. Allah be praised. Jerusalem, Tel Aviv, Haifa. My friend, I can't tell you how pleased I am. Israel will go 'poof'! They will cease to exist. The Israelis themselves admit they are a one-bomb country. So we will give them three!" He beamed, and allowed himself a single burst of laughter, then lowered his voice again, looking cautiously about the room.

"Leonide, as for our own capability... if you must know, we have modified our Block II Safir. It's a little over 70 feet long, about five feet in diameter. We've nearly doubled the payload capacity, now almost 190 pounds. Now our Safir is almost a replica of the Americans' Minuteman III, perhaps even a little larger. We can easily accommodate the warhead. Our missile technicians have affirmed they can manage the targeting mechanism. Good?"

"Yes, very good. And, I should not have to say this, but please use utmost caution, Bizhan. The risks here are exceedingly great. However, once successfully delivered to your hands and once employed, it will be a fait accompli, a 'done deal' as the Americans say."

"Of course, Leonide. Our operation is under a cloak of total secrecy. I must say though, this is huge. We will finally rid ourselves of the Jews. Praise to Allah."

"We will not miss the Jews either." Krasnoff smirked.

"We've let as many leave Russia as possible. You know, this will catch everyone by surprise. They'll be stunned. The world will wake up to a new reality."

"They'll wake to the smell of thermonuclear smoke! Indeed, the world will finally be forced to recognize the power of Iran. The ultimate beauty of this plan is the great uncertainty as to how many warheads we have. So unexpected. So sudden. What do you think is the time frame for delivery? We must close the window of vulnerability until actual deployment. The Americans constantly watch our launch pads with their damn satellites. We will announce this as simply another commercial satellite launch. And, once the Americans gather the telemetry data, it will be too late for anyone to do anything about it!"

"Bizhan, the warhead should be ready for retrieval and delivery to you by no later than December 1st. We will provide detailed instructions later regarding the actual manner of delivery."

Madani couldn't hide his joy. "December 1st? Yes, this will be a Hanukkah gift to the Jews! Fantastic. We will light their menorahs for them! Hah!"

"Agreed then?"

"Agreed. Leonide, one last thing. What about the CIA? Mossad? You know, they will do everything in their power to hunt you down and kill you for this."

Krasnoff snorted in derision. "In two months, Mossad won't even exist. Their headquarters outside Tel Aviv will be vaporized." He chuckled. "As for the CIA, I will be thousands of miles away with a new identity and plenty of money. My organization will be greatly appreciative as well. Bizhan, in truth, you will have more to worry about the CIA than I will."

"Oh, the Americans will do nothing. As we've said, it will be a 'done deal'... what can they do? They're weak and have no resolve, not anymore. The sun has set on the American empire.

No, we will remind those pigs as well as the whole world of Hiroshima and Nagasaki. Then of course, we will apologize and promise to everyone that we will never do such a thing again. There will be condemnation, attacks of words, more threats of sanctions. And that will be the end of it. And of course, the end of Israel."

"So, Iran does not think a nuclear attack on Israel would mean the immediate destruction of Tehran?"

"Again, Leonide, I ask you... by whom? The U.S. will not want to risk beginning a worldwide thermonuclear war. Russia won't lift a finger. So who?"

"I see. Hmm, okay, mon ami, we are in total agreement. *À bientôt.*" Leonide grinned.

The two men rose and shook hands, both smiling.

As Bizhan signaled, the bodyguard at the door glanced to Kasim and gestured toward the Russians at the nearby table. Kasim slid the large metal briefcase from under his table across the space between them to the waiting hands of the Russians.

Bizhan nodded to Kasim as he walked out of the Café Procope. The BMW sedan was waiting at the curb, the rear door already open.

Standing by the table, Leonide reached into his pocket for his wallet to pay the bill. And smiled.

Raphael raised his cell phone to his lips. "Dani, very good work. Look, you and Simone should leave now and return to the hotel. We'll contact you later. Pack everything up. All right, take off, Dani."

"Right." Dani pushed a now-frowning Simone from his lap, "Let's go."

Concern splashed across his face, Raphael turned to his partner. "Avrahm, this is not good. Something bad is brewing. I can smell it. Bizhan Madani and Leonide Krasnoff, Vevak and the Russian mafia conspiring? No, I don't like it. We need to get back to the Center for a post-op briefing. When's the flight to Ben Gurion?"

"7:00 PM this evening out of Charles de Gaulle airport. But Raphael, brief them on what? What do we really know?"

"We have a start. Considering the conditions here, Dani did some good work. It can help us to narrow the possibilities... what could these men and who they represent provide to the other. Can we move surveillance assets in closer on them to garner better intelligence? And if so, how to do it? Things like that." Raphael's face was filled with concern.

"Okay."

"Good. We have lots of time to make the flight. Let's go." The Audi swerved from the curb and sped down the Rue de l'Ancienne Comédie past the café, heading northeast through Paris.

CHAPTER 3

Anchorage, Alaska

McClure didn't dream on the five-plus hour flight. The stewardess in first class blessed him with three vodka cocktails, that and two Motrin© did the trick. Everybody else on the flight was napping anyway. His journey through the fog of sleep was broken only by the jolt of a quick loss in altitude. Startled, he looked at his wristwatch… 11:30 PM. A black night lurked outside the windows on their final approach into Anchorage. They were nearly an hour late, but it had been a pleasant, uneventful trip. Dave still couldn't figure why every damn nonstop from Denver to Anchorage had to be a night flight. No matter. His room and a nice fresh bed at the Marriott Courtyard would be waiting after he picked up the car at Hertz. McClure yawned and stretched his arms.

Driving out of the Hertz lot forty five minutes later, he found that the Grand Cherokee had a GPS. Nice touch, although he hoped he wouldn't need to use it. His breath fogged in front of the dash. Novak had been right about the weather here in November. At least two feet of snow had been plowed to the side of the road. He slipped his gloves on before pulling out of the parking spot, but it was just a short jaunt down International Airport Road, left on Spenard Road, and in minutes he was at the Marriott.

After registering, he grabbed two beers and wolfed down a chicken-salad sandwich from the Lobby Bistro before taking the elevator to the third floor. He threw his clothes in a pile on the only easy chair in the room. Hot water on his face felt wonderful. The fatigue of the five-hour plane flight, along with the alcohol, had soothed him enough for a sleep as seamless as the starless night outside. He plopped into bed and was gone in minutes.

McClure knew the drive to Valdez would be a long one, so he had to get on the road early. When the alarm jangled him awake at 5:30 AM, he slammed it off, and literally slid off the bed, blinking away his stupor in the stark darkness. He groaned, hunched over and stumbled Quasimodo-like into the bathroom. *Really do need to catch up on the zzzs in Valdez,* he thought.

Once showered, dressed and out the door, Dave hustled, a steaming cup of java in his hands. With the windows scraped free of frost and a light dusting of snow, the cherry red Grand Cherokee ground its way through the icy ruts out of the parking lot. The directions the desk receptionist gave him were easy really. Not much traffic. Many Alaskans adjusted to the short winter days by sleeping a little later on the mornings and working later into the inevitable early darkness. The gas tank was filled to the brim, so he motored out to 'C' Street. That turned into 'A' Street and after making a right onto 6th Avenue, before he knew it, he was on AK-1, Glenn Highway heading northeast to Glenallen. It'd be about four hours, maybe a tad less. He'd refuel there at the junction with Highway AK-4, Richardson Highway, before heading south to Valdez.

A frigid black night engulfed the landscape even though it was past 7:00 AM. A near-full moon was setting over the line of trees and seemed to be smiling at him. McClure smiled back. Moonlight on the snow produced a lot of light. For a minute, he could envision Anne sitting next to him, grinning and so awed at the sights. She slowly turned her head to him, her lips creased in smile. Almost real. He almost felt happy, first time in a year. Almost. But the vision faded as quickly as it came on, he shook off the fantasy and drilled his eyes through the windshield.

He whisked by an occasional house with its lights on, surprised to see that people chose to live so far out of Anchorage on this seemingly remote highway. The glare of his headlights didn't penetrate the dark forest that crowded the edge of the road, thicker than a box of toothpicks. There was no more than a foot of shoulder if you broke down. Thousands upon thousands of black spruce dominated the landscape, obscuring everything except the highway. Yeah, right now, he really *couldn't see the forest for the trees*, he chuckled.

A little after 9:00 AM, the sky was finally brightening to gray. The highway out the windshield was straight as the path of a rifle bullet. He was startled to see three caribou bound across the highway and back into the trees. They disappeared in a heartbeat. And every now and then, he saw a black spruce, not very tall but straight as an arrow, leaning over at an angle. He slowed a little to take a look. Black spruce had no real taproot, so in summer when the tundra thawed above the permafrost, the ground got soft and spongy, causing an occasional tree to lose its grip in the earth and bend over, sometimes at the strangest angle. Alaskans referred to these trees as *drunken spruce*. He remembered Anne reading it from a guidebook on their previous trip.

As the sky lit up, the sun now supplanting the moon, he slowed down, noticing a grocery store and the gas station just up ahead on his left. The very abbreviated community of Glenallen. It was 10:00 AM. McClure gassed the Cherokee at the station then pulled it into a parking spot in front of the convenience store. Soon, a cup of hot coffee steamed in his hands. In between sips, he slathered some relish and onions on a hot dog, and on the way to the register snatched a bag of Cheetos. Cheetos in the morning? Well, why the hell not?

It was near 10:30 AM when he pulled out and made the right turn onto Richardson Highway south to Valdez. It would be pretty much the same landscape until he passed through Keystone Canyon just northeast of town. Perhaps the scenery would be more open and prettier. He figured if he maintained his pace, he'd be approaching the harbor of Valdez by two o'clock in the afternoon.

He cracked the windows and air at almost thirty degrees Fahrenheit refreshed the cabin. McClure was wide awake with anticipation. Fatigue had left him, though he suspected he'd crash on a bed again tonight. The ride was gorgeous, but okay, a tad monotonous. Thick forests again, nothing but black spruce, the monotony broken only by the occasional leaning one. Snow everywhere. The Jeep made great time. Passing through the cliffs of Keystone Canyon, he slowed to about 35 mph just to take in the sights. Streams of water jetted down the stone face on both sides. Not a car on the road in either direction but the Jeep. Gorgeous. Mountains surrounded the town of Valdez like a giant bowl. He was soon smelling what he realized was the sea long before he saw it.

Suddenly, a road sign. Well, whaddya know, civilization! He passed Airport Road on his right. That's where he'd have to return the Cherokee to Hertz in the morning. Green forest still thronged along the road everywhere. Valdez was usually cloudy and foggy this time of year, but not today. A cobalt sky above, and looking up through the side window, he watched two seagulls pass overhead, patches of bright white against the blue.

McClure swung the Jeep left onto Meals Avenue and left again onto North Harbor Drive. The waterfront spanned to his right, the masts of fishing boats bobbing against the horizon.

Seagulls soared overhead, hoping for a bite to eat. Then there it was, the Harbor Inn. The snowy lot had been plowed, so he pulled right up front. He swept through the double glass doors trailing snow, and stamped his feet. After registering, he walked up the doublewide, carpeted stairway off the lobby to the second floor.

His room overlooked the harbor. Fantastic. The strong aroma of the sea was overwhelming. He threw the duffel bags and rifle case on the bed and hit the restroom. It had been a long drive and nature was calling. Once again, hot water on his face gave a much needed lift. After washing up, he changed his shirt and hopped back down the stairs to the lobby. The 'Off the Hook' restaurant located at the far end of the lobby had been hailed in his travel brochures as the best in Valdez. So, Off the Hook, it was.

The waiter watched McClure walk through the doorway. He turned, grinning ear to ear. "Help you, sir?"

"Got any Rose's Lime Juice?"

"Absolutely."

"Great. Then I'll have a Vodka Gimlet on the rocks, please."

"Bar stock?"

"No, Stoli if you have it."

"We do indeed. Coming right up, sir. Please, have a seat!"

Dave couldn't believe the view behind the bar. No mirror, just a fifty-foot-long view through a series of windows. The marina was brimming with boats and above them more seagulls than one could try to count. Mostly commercial fishing boats, but plenty of sport fishing craft, ready for hire. He imagined there were slim pickings for that in November.

The Vodka Gimlet was set on a coaster in front of him. McClure took a long sip. "Wow. Thanks. I just drove in from Anchorage."

"Really? Well hell, fella, you gotta be starved too. How about something to eat. Where'd you come from?"

"Colorado."

"Colorado? You like seafood?"

"Yesiree. It's only three o'clock though. Is it too early?"

"You staying here in the hotel?"

"Yes again."

"Well, this ain't called the Off the Hook for nothing. We serve anyone, anytime. How about some Alaskan king crab, baked potato, beans or Cole Slaw, fresh baked rolls, the works? The King Crab I'll give you is only off the boat four hours. Whaddya say?"

"Whoa. I say… bring it on!" McClure laughed out loud.

The waiter let out a belly laugh. "Yessiree! Coming right up." He turned toward the doors at the end of the bar. Dave stood, catching him in mid-stride.

"Say, where can I buy some ammo here, rifle, pistol?"

The waiter swerved back toward him. "The best place is only two blocks from here. Griz Outfitters. They got everything. Ammo, pack food, snares, everything. Going hunting?"

"Sort of. I'll be gone a week, snowshoeing. But I've got a rifle and a pistol. So, how do I get to Griz Outfitters?"

"Okay, out of the parking lot, turn left, going back toward Meals Avenue. Take a right on Meals, then not even a whole block up, go left on Galena. They're just a ways up on the right, just before you hit Tatitlek. Oh, and ask for Ed Weiss. The guy really knows the business and the environs around here. They open at nine… sun should be up by then. Although the weatherman says that tomorrow we're headed back into the fog and clouds. Way it is around here, that's all."

"Thanks again, fantastic. Thanks a lot. Say, where're you from anyway? What's the accent?"

"Norwegian. My parents came out here thirty years ago from Minnesota. White Bear Lake, just north of St. Paul. They loved it, so they stayed. I came out here about fifteen years ago. I liked it, so I stayed. I really don't miss Minnesota much."

"Norway, that's a beautiful country. Well, thanks for the info. I'll take a ride over to Griz Outfitters first thing in the morning."

"You're welcome, sir. A load of fresh King Crab for the hungry gentleman from Colorado coming right up!" He grinned and turned away, shoving open the door to the kitchen.

"I'll be ready for another gimlet by then too!" Dave called after him, and moved his glass to a table where he could still see the harbor. "Gees, this is great. Man, am I gonna sleep tonight or what?"

The masts and antennas of the boats in their slips swayed gently back and forth. It was a peaceful, quieting scene, almost mesmerizing. The estuarial waters of the bay, rising and falling with the tide, gave the boats a gentle roll. It had been a long trip, the plane flight and drive to Valdez, but now he was here and ready to jump off into the Chugach Mountains.

Slowly, Pete Novak's words began to crawl back into his head like the raw cut of a saw. *It'll be an icebox....you'll be knee-deep in snow. What if you can't shake it? What if you can't live off the land?* What if?

There was no 'what if'. No, tomorrow, he'd start the real journey. He raised the glass in a silent toast to the beauty of the harbor, took another long sip, and eased back in his chair with a contented sigh. "What tomorrow brings, tomorrow brings. Look out, Jeremiah Johnson, here I come."

CHAPTER 4

Tel Aviv, Israel

Raphael Mahler squeezed out of his five-year-old silver VW Passat and stood in the parking lot, his brown eyes turning west and sweeping the blue expanse of the Mediterranean. It was 60 degrees and beautiful out. A breeze ruffled his black curls as the sun warmed his tan face. Raphael had been born near Tel Aviv and then raised in a kibbutz far to the north at Kfar Giladi. After that, his parents decided to leave Israel.

His mother had written him last week from New York and so far, they liked it. They were discovering that the Jewish community in White Plains was an open and caring place. The move had been the right thing for them. Her sister was already there in Manhattan and helping them settle in. And his father, she wrote, was totally caught up in football and the upcoming playoffs, glued to the TV screen for hours every Sunday afternoon. Raphael smiled, thinking that if he had the arm of an American NFL quarterback, he could hit the sea from here with a stone. After all, the blue water of the Mediterranean was only just across the highway, a farmer's field and a short stretch of beach.

That thought brought him back. It reminded him of his father's oft-spoken and somber words, "We are such a small country, Raphael. Just over 100 miles wide. Always remember that a short-range missile fired east of the Jordan can fly over our heads and into the sea. So small, so vulnerable." Eventually, the constant stress of that possibility had pushed his parents to their decision to move to the States. Raphael looked up to the sky, imagining the sight of missile plumes trailing overhead. *Who would do such a thing to the peaceful people of Israel?* But he knew the answer: *Too many to count... way too many to count.*

For Raphael, a Sabra, a native-born Israeli, after college it seemed natural for him to join Metsada, the special operations department of Mossad. Sabras were held in somewhat higher esteem among Israelis and this applied to Raphael's Mossad counterparts as well. They appreciated his talents, and he appreciated the opportunity to contribute to the security of his country. His mentors had told him often that they were very impressed with his progress... he had a future with Mossad.

Since his surveillance team had returned from Paris six weeks earlier, everything had been quiet. Perhaps too quiet. And suddenly this morning, the call for a meeting at the Center. *Had to be Iran,* he thought. Iran's announcement to the world that she would allow inspectors at all her nuclear sites, that they could visit the country at will. Further, that Iran was in no way interested in the development of nuclear weapons. The world was stunned. Who would believe such a thing? He didn't, not for a minute.

He flung the light windbreaker over his shoulder and walked across the parking lot. As he approached the main doors of Mossad Center, just north of Tel Aviv, he thought, *Yes, just a stone's throw from the sea.* The headquarters, situated in a square grid of land bounded on the west by Highway 2 and to the south by Highway 5, was an impressive array of connected, boxlike buildings with center quadrangles. Beyond the roads, to the west, north and east, were agriculture fields making the structure even more prominently imposing in the landscape. Raphael glanced up at the windows above as he entered the building. He swept his proximity card through the turnstile as he slid through and punched in his pin.

The uniformed security guard glanced over to him. "Shalom, Raphael."

"Shalom, Ben. Good to see you."

He walked quickly to the elevator. It opened before he could press the button. Stepping inside, he reached over and punched '3', then inserted his proximity card. The window to the side of the buttons flashed 'Access Granted'. He could feel the elevator jerk upward. When the doors slid open, he stepped out and turned right.

The receptionist immediately to his left blinked her electric blue eyes and smiled at him. "Shalom, Rafi." She knew she was pretty, so she took liberties. She also knew he was still a bachelor, and her silky voice flirted with his name. "They're waiting for you in the Director's conference room."

"Shalom. I'm not late, Rebekah," he blurted, frowning.

"I know, Rafi, it's just that the others all came early," she smiled, once again batting her eyelashes.

He was not late. Still, he felt flushed and quickened his pace. Rebekah had said, "the others", so this meeting was not something routine. He walked through the open mahogany door and approached the long table. As he laid his jacket over an empty chair, he noted everyone was standing. He avoided direct eye contact.

"Shalom, Mr. Mahler, I'm glad you could make it," General Ariel Goren boomed. The general was broad-shouldered man with a very athletic build for his age. A dark tanned face, black hair and bushy black eyebrows. The steel in Ari's coal-black eyes had been tempered in the last 60 years of near-constant war. He was a two-star general in the Israel Defense Force as well as a deputy director in the Mossad hierarchy. General Goren had a 'take no prisoners' viewpoint when it came to Arab nations swearing to eliminate the nation of Israel. He strode to the front of the table, a screen display behind him. "Please, be seated."

"Shalom, General." Raphael slumped into his seat as the others pulled out chairs around the table.

As the screen lit up, Ari said, "This meeting is to bring you all up to date regarding our concerns for the apparent collaboration between Iranian Intelligence and the Russian mafia, specifically Bizhan Madani and Leonide Krasnoff. I trust you've received our assessment updates. By the way, Raphael, excellent work in Paris." Ari pointed to Raphael, then to the screen behind him. The photos of Madani and Krasnoff flashed on the screen.

"Thank you, sir." All faces at the table turned to him with nods of approval.

"Keep up the good work, Raphael, and you'll have my job someday." Muted chuckles followed Ari's glib pronouncement.

"First, I know I don't have to tell any of you that Iran's announcement this week that she is allowing inspectors to all her nuclear sites and disavowing all intentions of developing nuclear weapons is bullshit. Nothing but a deception, a smoke screen. If our friends in America can't see through the smoke, then we will have to do their seeing for them. The Americans haven't been all that friendly lately anyway." Laughter around the table.

"Now, all deception plans have a purpose. So it is our job to discover about what we are being deceived. Let's proceed." General Goren clicked the remote.

Two more photos flashed on the screen. Ari spoke again. "Kasim Shirazi. Shirazi works in the operations support department for Vevak." Another photo appeared on the screen. "And this large metal briefcase was carried by Shirazi into the Café Procope, then carried out by the Russians. This has been our primary concern."

Raphael was baffled. Where did that last photo come from? He interrupted, "Excuse me, General Goren, but… how did you obtain that photo of Shirazi and the case? Please. My team didn't…"

Ari waved his hand. "I'm sorry. Meira, please stand." An attractive woman in a light knit jacket and jeans stood up near the end of the table. Long brown hair, thirtyish. She turned her head and smiled at Raphael.

"Raphael, please meet Meira Dantzig. Meira is in Collections. She's proved her prowess in field operations on several occasions. We commissioned Meira's team on Kasim Shirazi. He arrived in Paris a day before Madani. He hand-carried the case with him. For security considerations, neither team knew of the existence of the other. If one of you were detained or otherwise compromised, your identities or affiliations exposed, well, we would still have had a second set of eyes in Paris. So, to both of you we say congratulations and thank you for your jobs so very well done. Your country is proud of you! Raphael, please, stand up also!" Ari clapped his hands vigorously and all around the table stood and joined in to honor them.

Both Meira and Raphael felt a rush of gooseflesh at the accolades. They grinned, nodding to all in appreciation, and eyed one another.

The image on the screen changed to a different view of the briefcase. "So… this case. We have since learned of a very large deposit, 25 million U.S. dollars, to a numbered account in a Zurich bank that we know belongs to the Russian mafia. Further, since everything has been so quiet since the Paris meeting and this transfer of funds, we suspect… well, that this is a down payment for something very big, immense, yet to come. Raphael, it fits with the number '25' that Dani assessed he saw Krasnoff say as well as the word 'deposit.' All of this is, of course, of grave concern."

A look of astonishment washed across the faces at the table. They glanced at each other in bewildered looks. Raphael looked to Meira. She raised her eyebrows, indicating she didn't have a clue as to what was transpiring.

"So our task, ladies and gentlemen, is to determine what the Russian mafia could possibly give to Vevak or do for Vevak that they couldn't do for themselves. Something that would garner such a price?"

Another photo flashed on the screen. "This is Major Vassily Krasnoff, GRU, Russian Military Intelligence. Vassily is Leonide's brother. We also know that Vassily is a very intellectually gifted young man and is on loan to the SVR, for what we don't know. Further, for reasons unknown to us, Major Krasnoff has been passed over for promotion to Lieutenant Colonel. We suspect it may have something to do with his brother being Russian Mafioso. Even stranger, the SVR assigned him to a Spetsnaz unit, Special Forces. This too is puzzling."

"Are you saying there's a connection to the business with the Iranians, Ari?" blurted Yakov Sivitz, Deputy Defense Minister for Strategic Systems, including extraterritorial actions, the sanctioning of assassinations.

"We don't know that, Yakov. At this point, we are only intrigued by the relationship… as well as by the basic question of whether Vassily would do something for his brother Leonide that was contrary to the Russian order, to his position, to his oath as an officer? So again, at this point, I just find it interesting."

The screen went blank. General Goren stepped to the side of the table, his face uneasy. "So, what to do? We are going to move in closer. Very close. I am assigning two new teams each to Leonide Krasnoff and Bizhan Madani. New teams, new faces. We will move in so close they will feel our breaths. They'll check the doors in their restrooms twice before they drop their pants. We hope such intense coverage may force a preemptive act, show their hand, provide us with leads for their plan, I don't know. At the same time, we will continue our collections and research, try to identify other relationships or business initiatives by the Russians and the Iranians."

"This spooks me, Ari. That's a lot of money, especially if it's only a down payment," deputy minister Sivitz interjected again.

"It does me too, Yakov. You all know me… I'm not an optimist. That's partly why I'm interested in Major Krasnoff."

"Ari, I agree. We cannot afford to wear blinders, to not look at every possible angle. But please, expand on your concerns for Major Krasnoff. You're holding something back." Ethan Kirschner, Assistant Deputy of Mossad for Collections, stood as he spoke. Meira smiled over at him.

For a moment, Ari stared at Ethan. "Old friends are often quite perceptive regarding each other, Ethan. Okay, we must look at everything, and also be sure that we're not chasing ghosts. But it's true that Major Krasnoff has me concerned."

"Ari, we all agree on that. Ethan's perception here may work for the better. Is it the brothers' relationship, the blood?" Yakov spoke softly.

"Blood, family… these things are strong, Yakov."

"Okay, but what else?" Yakov glanced to Kirschner. Raphael and Meira were mesmerized by the discussion.

The general continued. "It is difficult to collect inside Russia, this is a given. We do, however, have several low-level sources. Of course, they're not Jews. One is in the SVR's administration department. Very low-level."

"What's the communication method?" Yakov was curious.

"Burst transmitter. Encrypted of course."

"For such a low-level source?"

"Yakov, we're not exposing sensitive technology here. Everyone uses them, the Russians included."

Yakov nodded.

"What I was getting to is that four weeks ago, a Spetsnaz team including Krasnoff had orders to proceed to Murmansk. Lots of large crates, gear. We don't know what was in the crates."

"Murmansk? What's there?" Yakov again.

Goren hesitated. "Well, those orders were rescinded. The team received new orders and instead flew to Kamchatka. Petropavlovsk, the air station near the port."

Kirschner broke in. "Okay, what's the similarity between Murmansk and Petropavlovsk? Anyone?"

Several moments of silence engulfed the room. Everyone in the room glanced around in puzzlement. Then Raphael burst out.

"Subs!" He caught himself, embarrassed. "The submarine pens at Murmansk. And Petropavlovsk has a naval squadron that tends, ah, you know, services Russian subs coming out from under the ice and through the Bering Straits."

General Goren's face creased in a smile. "Good, Raphael. So, Russian subs go under the polar ice?"

"All the time, General. The Americans too."

"Excellent. So, assuming Major Krasnoff, the Spetsnaz team and all their gear is transferring to a sub off the Kamchatka Peninsula, where would they be going?"

Meira spoke. "Japan? North Korea?"

Ari puckered his lips. "No, I don't think so. The Russians could easily fly to North Korea. Why take a sub? And with Japan's increased maritime surveillance, the Russians couldn't get near the coast unnoticed if they wanted to. And why would they want to? No, we're missing something."

Several minutes of silence hung in the air like a choking fog. Again, their faces spoke bewilderment. Then Raphael once again looked up, locked eyes with General Goren and said softly, "Alaska."

Kirschner looked at him. "Alaska? What's in Alaska?"

"Eskimos," Meira smirked.

"Funny, Meira. We have a comedian." Raphael replied in monotone.

Meira scrunched her eyes, sighed and glared at Raphael.

Their eyes locked, Raphael returned a look that insinuated, *Well, lady, you started it.*

He continued. "The Americans have bases in the Aleutians... Shemya, other sites."

General Goren walked back to the front of the room. "And so? Okay, let's say Alaska is the target destination. But, Ethan is right. What's there? Why try to insert a team on U.S. Territory?" Ari suddenly angled his head to the ceiling. "Miki, do you have a map of Alaska back there? Can you bring it up on the screen? Quickly, please!"

A distant voice came through the speakers. "Yes sir, one minute please. Here it is." A map of Alaska filled the screen, edge to edge.

"It's huge." Yakov exclaimed. "Immense."

Ari broke in. "Yes, Alaska is gigantic. So, where to look? We do have a possibility here, but again, we must examine everything. I have to admit that the Alaska indications are intriguing. It's the "why" that baffles me though. We must consider this." Ari slid down into the chair at the head of the table and looked at each face in turn.

Raphael spoke, "General, what does the Prime Minister say? He knows of this, of course?"

"Oh yes, he does, Raphael. He believes as I do. The Iranians and the Russian mafia together could be a bad omen. Very bad."

"And?" Raphael's face flushed.

"And... I have been given the authority to be proactive in this matter. That means if this collaboration between the Russian mafia and Iranian Intelligence and the payoff in millions of dollars are things that will threaten our people, the nation of Israel, then we will… the only course is to terminate the threat. We will kill all those involved. Every one of them, like the Munich Olympics forty years ago. And further, we will kill all those who see us killing those involved. There will be no witnesses." He looked around the table. "I hope you understand our resolve to get to the bottom of this? Eh?"

"We do." In unison.

General Goren nodded. "Now then, thank you all for coming. Please, go back to your offices and put your thinking caps on. Don't over-focus on my musings regarding Major Krasnoff, Spetsnaz, and Alaska. Remember that is but one possibility. Also, don't do anything overt or provocative. Call me first with your ideas. We will meet again soon."

All stood. Some shook hands as they left the room.

"Raphael, Meira, please… sit down, stay." General Goren shuffled in his chair.

CHAPTER 5

Valdez, Alaska

McClure slept like a proverbial rock, dreamless. Once again, fatigue, vodka and another pair of Motrin© combined to carry him away to La-La Land in short order. After showering, he felt totally refreshed. He took the complimentary breakfast buffet in the lobby lounge. Damn good. Eggs, bacon, sausage, hash browns, toast and jam, all washed down with fresh coffee. The air in Valdez near the harbor was a brisk 18 degrees. The Jeep's windows were covered in frost and a light coat of snow. Hertz was thoughtful enough to include an ice scraper on the rear floor of the Jeep. So, after clearing the windows, he threw his gear in the Jeep and went back inside to the front desk to check out.

"Don't forget us! Come back soon, Mr. McClure!" Jennie at the counter chirped cheerfully.

"I just might do that." *Maybe, that is... if I come back at all.* That thought bumped him into the stark reality that his quest was about to commence. This morning, this day. No turning back now. Hours from now, in early November, he'd be alone in the Alaskan wilderness, trekking into the eastern Chugach Mountains and then along the southern edge of Wrangell-St. Elias National Park. The Wrangell-St. Elias was huge, 13.2 million acres, bigger than the country of Switzerland and exceedingly rugged terrain. This territory could be dangerous in winter, Pete Novak and others had either told Dave outright or implied it.

He leaned up against the driver-side door, jingling the keys in his hand. With vivid clarity he remembered reading Albert Camus in his college years. One particular quote of Camus had stuck with him through his life, especially so when years later he'd learned that a FBI colleague of his, dismayed over a recent divorce, had put a bullet in his own head. A real downer. But it drilled Camus' quote into his head. Camus had written, "But in the end one needs more courage to live than to kill himself." Or... to allow himself to die? Yes, he felt Camus was implying that as well. *So, Dave, what the hell are you doing? Are you really setting yourself up to end it here, to never come out of the Chugach wilderness? Are you by Camus' definition a coward to even think it?*

He climbed in the Jeep, slammed the door, turned the key and the engine roared to life. He jerked the wheel toward the street. *Horseshit,* he thought. No, this wasn't Dave McClure's personal death march, this was a quest. A quest to save his mind. It was a quest to learn how to cherish again the love he had with Anne and Michael. To hold them near his heart once more without breaking him in two. To thank God for the life they had together as a family.

The Jeep surged forward. McClure turned left out of the parking lot and took a right on Meals Avenue. At Galena Avenue, he paused, letting out a long exhale, then turned left and started down the street. There it was on the right, just like the bartender had told him. Griz Outfitters. It was 09:30 AM and his was the first vehicle to park in front of the store with its log-cabin facade.

A full-size carved wooden grizzly, baring its long claws and fanged teeth, stood on its hind legs ominously near the front door. "Grrr..." Dave growled as he passed by it. He walked up the several steps to the entrance, then swiveled in place, his blue eyes taking in the view of the surrounding mountains. The sun was on the horizon, peeking through a fog that was rapidly dissipating. He smiled, turned and pushed open the door.

"Morning, sir. You're an early riser. How can I help you? Nice parka you have on. Tree branch motif against a white background… perfect for the season." The voice came from behind a counter to his left.

"Hi. Thank you. I bought it in Colorado. Well, I need to pick up a few things, some ammo. I'm leaving in a little bit to go snowshoeing for a few days. Ah, can you tell me if Ed Weiss is in yet?"

"*Bin ich.*"

"What?"

"*Bin ich...* it's me. I'm Ed Weiss. Weiss is German for White. My family's old German stock. But really, I'm just Ed to everyone here in town. You too." His salt and pepper hair was ruffled, his face craggy, weathered lines across his forehead.

"Thanks. Well, hi again, Ed. Name's Dave McClure. You know, you come highly recommended by the bartender over at the Harbor Inn."

"Good to hear. Erik's a good guy. He's been here in Valdez quite a while too. So, tell me, what is it you're looking for? Whatever it is, we probably have it."

"Well, first thing is ammo. I've got a Weatherby .300 Magnum and a Smith and Wesson .44 magnum for a pistol, model 629 classic." Dave noticed another guy in the back of the store hanging packages on wall hooks.

"Okay, well, the rifle ammo's no problem. The Weatherby .300 Magnum has nice flat trajectory, long range and hits hard. You get 20 cartridges in a box. I don't know how long you'll be gone, but I recommend at least two boxes. Mind telling me what you have on it for a scope?"

"I'll take two boxes. Weight's an issue, you know. The scope's a Leupold Mark 4 4.5-14x50mm, illuminated Mil Dot."

"Wow, nice. With that baby, you can take shots out to 900, maybe even 1,000 yards. Fantastic. And, yep, two boxes of rifle ammo are always good when you're packing in." Ed paused, placing his hand on his chin.

"Mr. McClure, I don't know if you've ever fired it before, but for the revolver I have Winchester .44 Magnum platinum-tipped hollow point. It's really bad-ass ammo. Interested? There's 20 in a box."

"I've never seen or even heard of that ammo. You recommend it, eh?"

"Absolutely. That .44 round is unmatched for knock-down power, penetration, and tissue damage... 250 grains. The platinum plated wall on it maximizes penetration. And, it's notched around the mouth of the hollow point, which gives it tremendous expansion. Whaddya say?"

"Three boxes. You always want to have enough ammo. And, thanks for the tip."

"Also good. You have a flare gun? You know, just in case. They're very light weight, come with two flares."

"Yeah, I'll take one. I was thinking about it, but the airline wouldn't allow incendiary items on the plane coming out."

"Okay. Now, do you plan on being out there long enough to run out of your pack food? If so, what about wire for snares, building deadfalls...?"

"Boy, Ed, I'm sure glad you're here. Well, I thought that... I mean, if it was a survival issue, I'd drop a caribou or an elk. They're out there somewhere, aren't they?"

"Oh yeah, they're out there. If you're not actually going out to hunt them, we won't worry about a license. If it ends up being a survival issue, nobody would give you any static about that. There's a big herd of caribou, the Nelchina herd. Last time

I heard, there was over 46,000 caribou in that herd. Problem is, the area is huge so you can't ever be certain where they're gonna be. It's very possible that you might never see one. And, there's no elk up here. They put some out on Etolin and Afognak Islands to see if a herd would get started. But, not here though. There'd be too much competition for food."

"I see. Any moose?"

"Yes, moose are out there. Not in great numbers though, and they're a little reclusive. Anyway, that's why I mention the snares. Oh, if you should see caribou and a couple approach you, unless they're calving, there's no real danger. Caribou are curious as hell. If they're unsure of you, what kind of creature you are, they get nosy and come to investigate you. That's all. Kind of interesting, really. And the females are the only ones in the deer family that have antlers. They don't lose them until May or June after they calf. The males will start losing theirs this month and in December. All the caribou start growing them again in August. Just a nature lover's tidbit for you."

"That's all very interesting. I didn't know any of it. Again, I'm very glad I ran into you, Ed."

"Same here, Dave. I always tell people that there are millions of rabbits out there, and they're easy to catch if you know what you're doing. Pretty tasty grilled on a spit too. Can you build a simple spring pole snare, some call it a twitch-up snare?"

"Yes, I know how to do that. Training long ago. I don't have any wire though."

"I can take care of that for you too... high-tensile strength aircraft cable. I'll give you about 30 feet of it. It works great. Use about twelve inch loops. That'll catch anything from a rabbit up to a wolf. Oh yeah, and you should know a pretty good sized pack of wolves is operating in the eastern Chugach south of the Wrangell-Elias. They've marked off that whole area as their territory. I've never heard of any hunter having a serious run-in with them, though. And, a few creeks are still running for a little fishing too, if you're interested."

"Thanks again. I'll keep that all in mind. Any bears you know of? They're hibernating for the winter by now, right?"

"Ah, most of them, but not all of them. There's a pair of males out there that never hibernate the whole winter. My brother-in-law, Jack Stroud, he says he sees those two bears almost every winter. He's named them Bruno and Brutus, though I don't know how the hell he can tell them apart. Anyway, best be on alert for bear sign. Jack's out there now. He goes out for weeks at a time. I doubt that in a wilderness that large you'll see him, but if you do, please tell him that his sister and I would like to have him over for dinner two Sundays from now. Jack's a little whacky, keeps to himself, and he drinks but only when he's out in the forest. But basically, he's an okay guy. Like I said though, I doubt you'll see him."

"That's good info on the grizzlies. Bruno and Brutus, huh? Well, if I see them, I'll have a little chat, let them know that they have names now. I'll keep an eye out for your brother-in-law and anybody else for that matter."

"You'll chat with the bears? I'm glad you have a sense of humor, Dave, because you may find you need it out there." Weiss paused, giving McClure a serious look. "And oh, do you have a pack sterno, or any other way to always be sure you have a fire?"

"Ed, I have a pack sterno with three cans. Plus, I've got a magnesium block for putting shavings on kindling. They say magnesium shavings burn at up to 3,000 degrees. Also, four packs of windproof, storm matches. They're supposed to start even in real high winds."

"Well, you're in great shape there. I assume you already have pack food for a few days."

"Yes, I do."

"And you can melt snow for water. I think that's about it. I'll bundle this up for you. We take all major credit cards." Ed grouped all the supplies at the counter and packed everything in a bag.

"Other than that, Dave, I'd just say it's always good to remember how much daylight you have left. That way, you can plan to start a camp in the light. Right now, it starts getting light about 8-9:00 AM and dark again about 3-4:00 PM. Temps in the Chugach are cooler than here, so it ranges from about 0 degrees to 30 or so. Oh, and I'm gonna give you a color brochure on plants, the poisonous ones. Believe it or not, Alaska has some real humdingers up here. If push comes to shove and you go digging for roots, frozen plants under the snow, this will keep you from making a fatal mistake."

"All good info."

"By the way, have you decided where you're gonna go in? When're you planning to start, and how're you getting there? "

"Well, the Jeep out front is a rental. After I leave here, I'm gonna drop it off at Hertz at the airport. I guess I'll ask them for a lift as soon as I add this stuff to my pack. I'm going in this morning."

Ed nodded. "Tell you what. I'll follow you up to the airstrip, it's only a couple miles. Then I'll take you where you want. If you don't have a starting point already picked out, the best place is just north of Keystone Canyon on Highway 4. You passed it when you drove down here. There's actually a little trailhead. It runs for a few miles, then trickles off to nothing. Just stay west of Brown's Creek for as long as you can. You'll see it. Then, when the creek dwindles as it flows north, cross over on the small bridge, not on the creek. Ice isn't thick enough yet. One way or another, if you're heading east, you'll be crossing the Copper River. It's pretty narrow at some points. After that, head north, northeast as you find you're able to, then directly east."

"Gees, that's a great offer, Ed. I don't know how I can thank you."

"Well, I'd like to know where you went in, anyway. Just for safety purposes, you know. Say, do you have a cell phone?"

"Yes."

"Well, there won't be any coverage when you get a day's trek into the wilderness, but why don't you give me your number and I'll give you mine. You can call me when you come out, hit the highway, and I can come pick you up. Bring you over to the hotel or my place, whichever you prefer."

"Wow, another great offer, Ed." They exchanged cell phone numbers. McClure scribbled his on a scrap of paper, while Ed wrote his on the store receipt.

"Good. Hey, Hank..." Ed shouted to the rear of the store. "I'm gonna follow Mr. McClure here out to the airport and then drop him off at the canyon trailhead. About an hour is all. Take the register, okay?"

"Got it, Ed!" Hank strode down the aisle to the front.

The ride to the airstrip was a quick jaunt. After sliding the keys across the counter to the Hertz agent and signing the credit card receipt, McClure climbed into the front of Ed's gray Yukon. They swung around the ice coated parking lot and then down the hill from the air strip to the Richardson Highway. The road was clear, and they were at the north end of the canyon in no time. Ed turned into the small trailhead entrance past the shoulder, shifted into park, and turned off the engine.

"Need some help?"

"No, I think I got it. Thanks for everything, Ed. I'll see you when I come out." He reached for the door handle.

"Any idea when that's gonna be? Ah, Dave, if you'll allow me, I'd like to ask you... why are you doing this in November? It'll be tough in there. The winter in the Wrangell-St. Elias can be very unfriendly. Frigid. How come now?"

McClure turned his head and their eyes connected. "It's something that's deep inside me, Ed, something I have to deal with, come to grips with. I have to do this. I have to clear my head, tuck away some bad memories. If I don't, I think I'm gonna go nuts. Going in here now is just what the doctor ordered. I just have a feeling it's right for me."

"All right. Well, thanks for sharing. I understand. In some ways, it's kind of like why my brother-in-law Jack treks in there every winter. It helps him somehow, I guess. I'll tell you what, I don't make friends easily, Dave. Please get through this. Come out. I'll buy you a drink and we'll have dinner at my place. Hell, sounds like you're in no rush anyway. Deal?"

"Deal. I'll look forward to it. Thanks a bunch. Gees, it's almost 11:00 AM, I gotta get to it."

Ed watched as McClure slid out of the Yukon and opened the back door for his pack, rifle and snowshoes. He walked off the shoulder into the snow, and waved.

Ed smiled through the glass and waved back. The Yukon pulled out, did a U-turn, and disappeared down the highway. Dave turned and stomped over the tree line, dropped the pack, sat on it, then strapped on his snowshoes. After pulling on the backpack and slinging the rifle over his shoulder, he took a deep breath of mountain air. He telescoped his two snowshoeing sticks, pushed his goggles down over his eyes and scanned the snow-covered trail that led uphill from the trailhead. *Well, here it is,* he mused. *Game time.*

CHAPTER 6

Chugach Mountains, Alaska

The trail was on a steady rise for quite a distance and the going was tough. McClure plodded ahead, his throat growing raw from constantly sucking in breaths heavy with frost. But in an hour or so, as the sun rose in the sky, it took some of the ice out of the air. Much better on the lungs. McClure got into the rhythm, an easy gait of planting the snowshoes and hiking sticks, one after another, up the trail. His breathing settled into the same pace, in - out, in - out. There wasn't that much snow yet, less than two feet, and the snowshoes kept him from sinking more than four inches.

He watched the sky through the trees as he advanced on the trail and saw the clouds breaking up. The sun was peeking through. Off in a corner of his mind, he knew the late start would give him a short day today, and remembered he shouldn't wait too long to start a campsite. As the trail began to level off, he made a habit of frequently scanning right and left of the trail for animal sign: hoofs, bear tracks, wolves, whatever. Nothing yet. Some faint depressions indicating snowshoes leading ahead of him, but they were barely noticeable, having been filled in with snow. So, not a single soul was out on the trail today, or yesterday for that matter, but him.

Minutes passed into hours. It wasn't long before the ache in his legs reminded him that he hadn't used this particular set of muscles in a long time. He hadn't been snowshoeing in years, so he slowed a little to keep from cramping up. Eventually, he broke through a tree line to find Brown's Creek dead ahead. Lots of rocks and pebbles. Brown's Creek really wasn't much of a creek at all, not now. He imagined in the spring though, the waters would swell their banks to nourish the trees and vegetation to the south. Even though it didn't appear to be that long a distance, he sat down and unstrapped the snowshoes. The last thing he wanted to do was bust a shoe on a rock.

He looked north and south as he passed over the small bridge. The creek meandered away in both directions, trees reaching out on both sides. A fallen tree lay by the wooded edge of the creek's shore. McClure sat down, slipped his boot into the toe hold once again, and drew the straps of the snowshoes tight. He picked up the trail again, eagerly plodding ahead while keeping an easterly path.

Dave was now entering the Chugach wilderness in its full glory, skirting the southern boundary of the Wrangell-St. Elias National Park. McClure paused, lifting his nose to the air. He hadn't smelled anything so intensely fresh in years. The rich scents of spruce, birch and fresh snow were invigorating. He also noticed that the tree density was steadily increasing as well, closer spacing and packing up with a wide variety of white spruce, black spruce, pine, birch and balsam poplar with occasional patches of aspen. The forest was everything he had expected and more, simply magnificent. He also noticed that the once-clear trail ahead was growing ever more narrow, dwindling off to nothing. Ed Weiss had told him to expect this. It would be a strange feeling at first, the forest closing in about him, almost hugging him. He continued to tramp forward, remaining on a generally easterly vector as the woods ahead grew tighter.

He soon entered a stand of birch, pretty tightly packed except for a little break off to his right. The clearing was smallish but just right for his purpose. The sky was graying. McClure decided this was it. He'd camp here, take it easy for the rest of the day, get a good night's sleep and press hard ahead tomorrow. He sensed a faint trace of water in the air. The Copper River had to be just a short distance ahead anyway, and it might take him a while to find a good crossing point. He knew he was way to the south of the Chitina River and the 60-mile-long gravel road out to McCarthy that cut through the Wrangell-St. Elias, so the going for him wasn't going to improve significantly. Yes, this would be his first camp site.

He slid the Weatherby rifle off his shoulder and leaned it against a birch, then dropped his pack to the snow with a small sigh of relief. He plunked down on the pack, tracking his eyes right to left, just taking in the sights. On his left wrist, his watch read... 3:00 PM. Light was diminishing. Water from the bottle he'd filled at the hotel felt good, soothing his parched throat. The air was chilly but not overly so, and he noticed how incredibly still the air was, not even a whisper of a breeze. The woods here were just too thick for any strong wind.

Dave stood, his snowshoes still strapped to his feet, and began walking in a circle around the small clearing to pack down the snow. He pulled out his SOG titanium knife, trudged over to a pine, and began cutting some boughs for a mattress. Once he felt he had enough laid down in a long rectangle on the snow, he unstrapped his snowshoes, leaned them against a birch and rolled out the one-man tent. The pack tent was nylon and small, only about three feet tall at its peak and four feet wide, but that was more than enough. The tent sprung up easily. Inside, he rolled out the thermal sleeping bag to its full six and a half foot length. He slid the pack in the side of the tent, then the rifle. His .44 magnum revolver was still strapped to a belt holster under the parka.

The snow wasn't too deep, only about a foot of the white stuff in the little clearing. He scooped out some for a fire pit. After gathering some dry brush and branches, he chopped a few larger branches with the hatchet and stacked them off to the side of his fire pit. He tee-peed the brush and dry branches, then pulled out the magnesium block and the windproof matches. Squatting in the snow, McClure hummed an obscure tune as he methodically scraped magnesium shavings into the brush. Once he struck the match, the bright flame accentuated the dwindling light in the surrounding forest. The brush flared up and the fire started immediately. The magnesium shavings easily did the trick burning to a high of three thousand degrees. He added larger branches. Last, he scooped some snow into the pot for water for his freeze-dried vegetable beef soup and coffee. He sat back on some spruce boughs in front of the tent opening and smiled. This was great. Just one more task.

McClure reached in the tent for his pack and pulled out the coil of white nylon rope with the metal stake attached. He unhooked the small camp shovel from the pack and laid it inside the circle of the campsite. Pausing to add some more snow to the simmering pot, he stood and stepped to the perimeter of his clearing. He wrapped and tied off one end of the 60 foot rope around a birch trunk just behind the tent. Then, walking outside the edge of the clearing, he drew the rope taut, making a circle about two feet above the snow and twisting the stake on the rope tight as he went along until he returned to where he had already first knotted the rope.

Dave cut a notch in the trunk of the birch and gently edged the tip of the stake against the notch, then looped the loose end of the rope around the birch. He dropped to his knees and slid under the circle of rope to retrieve the camp shovel. Finally, he pressed the shovel into the snow a few inches from the metal stake. His perimeter alarm was now set. If the circle of tightly twisted rope were bumped or somehow tripped during the night, the rope would snap the stake loose, spinning it and slapping it repeatedly against the shovel. Lots of noise.

Night soon swallowed up the wilderness. The birches along the perimeter grew into dark sentinels against the backdrop of snow. The moon skimmed along the horizon, sending shadows of birch trees creeping across the snow. His eyes opened wide as far in the distance, the sudden, singular howl of a wolf pierced the silence. Two more calls from different wolves followed, reverberating along the valley. The wolf pack Ed had mentioned. Although he had no real way to tell, the howls sounded very far away.

The fire was burning well, providing light and warmth. It drew the focus of his attention as the details of the surrounding forest melted into dark shades of gray. The soup was excellent, the broth thicker and tastier than he had imagined it would be. A hot cup of coffee with a smidgeon of sugar finished dinner and felt good in his gut. He cleaned the pot and spoons in the snow and covered the area with six inches of fresh snow to hide any smells. McClure added some large branches to the fire, sending sparks soaring up to a night sky now teeming with twinkling stars.

He sat in front of the tent considering the woods. Scraggly arms of leafless birches were packed closely but still allowed him some view to the distant north. The snowy Chugach mountains were illuminated by the silver ghost of a moon still sliding across the horizon. It would set soon. The scene was one of such quiet splendor that he knew he would have difficulty describing it later to someone. Total stillness.

A dull "pop" sounded straight ahead. Another. Temperatures must be dropping precipitously. The sap froze in some distant branch, popping the bark. McClure was so caught up in the majesty of the surroundings that he'd been oblivious to the rapidly dropping temperature of the night. He added a couple more large branches to the fire and crawled in the tent. In the tight surroundings, he struggled a bit to slip off his parka and boots, but finally shoved them against the tent wall and zipped up the thermal sleeping bag around him. He took the revolver out of his holster and pushed it up by his chest within easy reach. He zipped the tunneled flap of the tent about halfway down, allowing him some vision of the campsite's clearing.

It was only nearing 6:00 PM but everything was now the darkest of grays. Not long from now, the moon would finally slip below the horizon and the Alaskan night would overwhelm the wilderness. The sleeping bag was rated down to -20° F and it was doing a great job thus far, keeping him snug and warm. Dave was fatigued, but fought against closing his eyes. He preferred to drift off into a soothing blackness void of the dreams of REM sleep.

Dave had a bottle of Motrin but didn't want to dull his senses completely. Even though he'd seen no signs of forest creatures, he had heard the wolves. So, it was within the realm of possibility to have a nocturnal visitor to the camp. He'd made his perimeter secure and had the .44 magnum within reach, making him as safe as reasonably possible in the Chugach mountains.

Without warning, a loud clapping sound caused him to poke his head out. There it was, directly in front of him in the middle branches of a birch on the perimeter circle. Although it wasn't clapping, it was flapping to give its wings a stretch. A large Alaskan snowy owl, a big one, almost two feet tall. It swiveled its head toward him and McClure could feel the dark eyes staring into his own, not threatening, just curious. A magnificent bird.

"Say, would you mind keeping an eye on the place tonight?" He spoke softly.

With a single flap of its wings, the owl blinked, settling down into a ball of feathers on the branch, keeping an eye on him.

"Was that a yes or a no? Just watch things for a few hours. Whaddya say?" he chuckled.

The owl blinked again.

"Okay then. 'Night." A wondrous glimpse of a snowy owl not twenty feet away. He shook his head in amazement. McClure tucked his head back in the tent and re-zipped the front flap. That was truly a once in a lifetime sight.

Dave wondered what Anne would think of him being out here now, alone in the Alaskan wilderness. She probably wouldn't believe it. Oh, they'd loved hiking together, but their idea of camping out had been restricted to five star resorts at the edge of the wilderness. The forest was always within easy reach, but never totally enveloped them. No, Anne wouldn't believe it.

He pushed the side of his face into the soft lining of the sleeping bag. Anne and Michael were so close in his thoughts, but so distant in his reality. He wished he could reach out and touch her face, draw her close. He wished he could kiss her. He wished he could feel the press of her breasts against his chest. And, he wished he could look into Michael's blue eyes and give him a hug that would last him a lifetime. He wished he could gather both of them in his arms and tell them how much he loved them. He wished.

CHAPTER 7

Jerusalem, Israel

Raphael sipped his coffee in the shade of an umbrella outside the Babette Café. The street was already busy with shoppers. The latest food rage in Israel was waffles of all sorts, but Belgian waffles mostly. The Babette Café in Jerusalem's city center was very small, tiny as a closet with just five tables inside and about as many tables with umbrellas scattered on the walkway outside. Easy to walk past and miss. But you couldn't beat the waffles anywhere in Israel. Babette's on Shammai Street just north of Independence Park, was applauded as the best of the best, offering sauces from chocolate and chestnut cream to applesauce and other toppings that would make one's eyes roll up in pleasure.

The sun was out, and at 10:00 AM it was already over 60 degrees. Raphael's eyebrows rose slightly as he watched Meira turn the corner not quite a block away. Her hips swayed gently as she walked. He wondered if women practiced such a thing. They all seemed to do it, swivel their hips slowly back and forth when they walked. It caught his eye anyway. He took another sip of coffee. She was wearing a light blue sweatshirt with the sleeves pushed up, jeans and tennis shoes. Her long brown hair bounced softly around her shoulders. Her lips quickly broke in smile when she noticed him.

Okay, he admitted it to himself… Meira was good looking. He smiled back. He hoped she liked waffles. He rose and pulled out her chair for her.

"Thank you." She said, easing into it.

"Good morning, Ms. Dantzig," Raphael said stiffly.

"So formal, Raphael. Please, may I call you Rafi?"

"We really don't know each other very well, Meira, but sure, Rafi is okay with me. That makes two of you, you and Rebekah."

Her eyes opened wide, then she spoke firmly. "Don't compare me with Rebekah, Raphael. There is no comparison. I don't jump into bed with a man because I think he might help get me a raise or a promotion. If you want Raphael, so be it."

Ouch. He winced mentally. "Ah, Meira, I'm sorry. I wasn't making a comparison. In fact, Rebekah does grind on the nerves with the liberties she thinks she can take as a secretary."

"Okay, well, I'm sorry I jumped on you. I just don't care for Rebekah or for her type. By the way, nice café. Are we here for waffles with strawberries and whipped cream?"

"If you like them, yes."

"I love them. It's a long drive for waffles though, isn't it?"

"On Highway 1, the *Kvish Ahat,* it's under 40 miles from Tel Aviv, less than an hour's drive. I assume you came that way. And, it's hard to beat Babette's for waffles." Raphael waved to a waiter. They both ordered Belgian waffles. Meira also asked for coffee. Two plates of light, crisp waffles with strawberries and whipped cream soon adorned the table. Raphael watched Meira with curiosity. He found it interesting the way she ate or rather devoured, her waffles. First, all the sliced strawberries disappeared in gulps, then she slurped spoonfuls of whipped cream after plopping a dab in her coffee. Finally, she attacked the waffles with the berry juice and remainder of the cream.

"What?" she asked, realizing his eyes were on her.

"Nothing. I was just noticing how you divided up your

breakfast into groups for a gastronomic conquest."

She slipped into a smile. "You're not the first to say so. I love strawberries so much I can't wait for just bite-sized portions. So, really, Rafi, why did you choose to meet here, an hour away in Jerusalem?"

"I just really love it here. I come here often. Sometimes I like to just walk by the park, sit and think. I walk along these weathered streets in this centuries-old city, the heart of Israel. Jerusalem gives me focus. Then of course, there's the coffee and waffles." He laughed.

"Jerusalem gives you focus. I think I can understand that. There's more to you than meets the eye, Rafi. Although you are easy on the eyes." Her eyes rose to meet his.

He smiled in return. Raphael found that comment interesting. He felt a stirring.

"Truth is, I suppose I should have assumed so with you being a Metsada team chief." Meira continued. "You must be performing extremely well for General Goren to make a comment about you taking his job someday. So please know, Rafi, I am happy to be working with you. Special operations seemed a little scary to me when I graduated from the basic school, so I chose Collections."

"Oh, Ari was joking. He was just being flippant, trying to break the ice. My team did well in Paris, but so did yours."

"The general doesn't make such remarks gratuitously, Rafi. Anyway, thanks for inviting me here for breakfast."

"My pleasure. Collections is no cakewalk either, Meira. You were also leading a surveillance team, and successfully so. I am happy to be working with you too."

"Good. Then, this is a wonderful start. We're both just so very happy to be working with each other. So, now what?" She smirked.

That was a quick change in demeanor. *Why?* Raphael

thought, but didn't let his surprise show. Without skipping a beat, he instead replied promptly. "Now? Now we discuss the Paris meeting: what we gleaned from it as well as our understanding of the Russian mafia and Vevak. What's possible? Specifically, what could be going on that's worth a $25 million down payment? What can Krasnoff do for Madani that he and Vevak could not do by themselves? That's what's now. That is, if that's not too difficult." He paused, scanning her face.

Raphael continued. "Meira, I don't understand your tone. If you don't want in this, say so now. You confirmed your participation at the meeting with General Goren. If you've changed your mind, say so. I'll find someone else." He stared at her, his jaw set, and purposely dropped his fork, letting it fall on the white china plate with a clatter.

Meira blinked, her eyebrows rising, cheeks flushing. Rafi was good on his feet. Intelligent and intuitive. He could be aggressive. And after all, she knew she was being a snot. "Rafi, look, I'm sorry I jumped again. My bad. Really. Look, I do want in this, whatever it turns out to be, and I recognize your leadership. Please, give me another chance. Let's start over."

"Okay then. Thanks." He relaxed his shoulders.

"Thank you, Rafi. You know I do like the guy I'm having breakfast with. You're a nice guy and you've got a lot on the go. I have a good gut feeling about you. The last thing I want to do is build walls."

"All right. Well, maybe I should be a little open with you so you'll understand me better. Fact is, Meira, I myself am no stranger to building walls, especially with the opposite sex."

"Are you divorced?"

"No. But I was engaged."

"Engaged? Still hurt?" Meira leaned forward, setting her elbows on the table.

"It tore my guts out. I read her so completely wrong, and

paid for it in spades. I guess… well, it's one thing to be rejected after falling so hard for someone, but it's quite another thing to be played, manipulated and then tossed away like a piece of trash. She was sleeping with another guy all along."

"The bitch. That's being a whore. Look, Rafi, I understand." Her voice dropped to a whisper. "Thank you so much for opening up and sharing that with me. So, are you saying you've blocked any thoughts of another relationship?"

"For now, I would say yes. I'd have a hard time trusting someone, trusting them fully. I suppose it's a self-defense mechanism. Meira, you're the first one I've told, in fact, it's hard to believe I've said this to a woman."

"Yeah, shame on you." She chuckled. Her right hand slid across the table and covered his. "Buddies?"

"Yeah, buddies." He raised his eyes. They were misty. "And so, since we're sharing, what about Meira?"

Her eyes grew soft, her voice lowering. "Meira is still looking for the right man. She's in no rush. I try not to have walls." He noticed she had hazel eyes. Nice eyes.

"Good."

"Okay then, Rafi, General Goren told us to take the lead in this, to put our thinking caps on. What the hell could be worth 25 million? And as a down payment yet? An assassination… our prime minister? An American senator? A president? Perhaps another 9/11 bombing, but more massive?"

Raphael leaned back in his chair. "I suspect that if any of those things were to occur and were in any way traceable back to Iran, full scale military operations would commence immediately. Iran would face tremendous destruction. The revenge would be so severe, so total, that the Iran that exists today wouldn't be around to pay the balance to Krasnoff and his Russian mafia bosses."

"I agree, Rafi. It's logical. What then?"

"Well, I think we're all ignoring the obvious. We're all afraid to say the word, so we stick our heads in the sand and assess other possibilities."

"What?"

"Meira, what do we fear most? What could be the absolute very worst, something so bad the nation of Israel could not withstand, not survive? And, what would be worth it to Vevak, to the Iranians, to pay $25 million, $100 million or even more?"

"You're scaring me, Rafi." In fear, she pulled her hands back from the table and clenched them in her lap, shaking her head slowly back and forth.

"Say it."

"A nuclear weapon."

"Right. That would certainly fit with the sudden, earth-shattering media announcement by Iran that she will allow inspectors at all her nuclear sites, wouldn't it? That she disavows all intentions of developing nuclear weapons? Deception. A classic psychological diversion as a precursor to a nuclear attack."

"My God. Do you think...?"

"General Goren even said the announcement was bullshit. Everyone at the table agreed. But no one wants to admit the obvious, the most evil thing that we could face as a nation. Instead, we stick our heads in the sand."

"But where would Vevak, or actually the Russian mafia get one? Who has them - I have to strain my brain here... the U.S., Russia, the U.K., France, China, India, Pakistan, North Korea... and us. Yes?"

"Yes. Well, I think we can safely disregard us. Russia also, her nukes are simply too big, too unwieldy for Iran's Safir missiles. The United Kingdom? Ah, too small a force and all locked up like Fort Knox. I don't think the RAF even trains with her own nukes. France? Even smaller, a handful, all tucked away, not out and not operational. So, what's the possibility of China working on a deal with the Russian mafia? Forget it. With the state of relations between China and Russia, how could that happen even with the Russian mafia? When China is trying to establish her position as a player on the world stage, she has too much to lose.

India and Pakistan? They hate each other but both are talking with Russia. India is even working together militarily with Russia, especially with her navy. Perhaps Russia working with both hostile nations is good for stability. Finally then, North Korea? The Russian mafia would find making overtures to North Korea impossible. Their emperor-like leader is too paranoid to buy anything like that. If either Krasnoff or Madani showed up on North Korean soil, they'd be executed as spies. So, who else?"

"The United States. They have a large, mixed nuclear force… bombers, missiles, submarines. But, I can't imagine anyone getting close enough. The U.S. is too big, too powerful, too well protected."

"Rome once thought so too. We know what happened there. The Americans have had some problems with their nuclear surety program. It's been in the news."

"And you think that's enough, a recent problem in the news? That makes the U.S. nuclear force the only real target?" Meira focused on his eyes. She suspected he was holding back, something in the puzzle he had pieced together that pointed to the U.S. "Rafi, what is it? What aren't you telling me?"

"You are very intuitive, Meira. I like that. Well, it's a couple of things that Dani said he thinks he lip-read that Krasnoff said to Madani. We can't ask Dani about it now. It would serve no purpose… whatever he thinks he heard, he heard. Why don't we finish up and go back to the Center? I would like to talk to a couple of your people in Collections, techies who have a handle on other nations' nuclear weapons: their yield, types, warhead sizes and weights, specifically the United States."

"Okay, but what was it that Dani said? Tell me."

"The number '450' and the suspected additional name, 'Marv' or 'Merv'. I think that's what he said. It's something in the back of my mind that I remember from intelligence school."

"Intelligence school? I wish I knew where you were going with this. It makes me nervous. I think..." Meira's eyes slowly drifted to her right and her head turned. Her focus was beyond Raphael's peripheral vision, on something he couldn't see.

Meira gradually rose from her seat, reaching under her sweatshirt to the belt holster. Unsnapping the retainer, she drew her Walther PPK pistol up bit by bit, cleared the holster, then lowered it alongside her right thigh. Instinctively, her right thumb pushed up the lever to take the safety off. Meira edged away from the table slowly but deliberately.

"No, noo," she muttered.

"What? Meira..." Startled, Raphael swiveled in his chair but couldn't see whatever she had locked on to.

Meira stepped from the table and walked toward a fabric store just up the street on the right. She drilled her eyes into a man in a tattered tweed sport coat and baggy pants. Dark complexion, beard. His eyes skittered quickly from side to side. He appeared nervous, even afraid. Beads of sweat stood out on his forehead. He stopped in the doorway and set down against the wall an oversized leather bag he'd been carrying. He then began to step away to the street.

"P'tzatza! P'tzatza!" 'Bomb! Bomb!' Meira shouted. She leapt forward as if from a starting block and sprinted toward the man. Raphael stood up, hesitant. What? He didn't yet see it. People on the street instantly dove for the pavement and placed their hands over their heads. Mothers pushed their children under them. It was a conditioned response in Israel.

The man's head swung left to Meira. His eyes widened and

his hand moved to his side, reaching for the flap on his sport coat. Seeing Meira's collision with him was imminent, he instead reached further down, pushing aside his jacket and grabbing the grip of a Colt .45 that protruded from his belt.

"*Yahud, Anti Khanzeer!*" "Jew, you are a pig!" he shrieked at the top of his lungs.

Meira brought the full force of her left arm forward, punching him square in the face. Again. And again. The man stumbled backward and slid down the wall, his hand still on the pistol grip. Meira fell with him. The man grabbed her by the throat and squeezed with everything he had. His finger curled around the trigger of the Colt, but the barrel wasn't clear of his belt. A loud blast of sound and flash echoed down the street. He screamed as the bullet ripped down his thigh, gouging his kneecap.

Meira's lungs begged for air. She gasped, repeatedly thumping the man in his temple with the Walther pistol.

"Arghh!" He grunted, struggling to free the pistol's barrel from his belt. Another blast. The round missed Meira, went under her and ricocheted off the concrete. An old man howled behind her and grabbed his leg in pain.

The man suddenly released the grip of the pistol and fumbled desperately for his jacket flap pocket. Meira shoved the Walther pistol under his jaw. "Don't do it!" she shouted into his face.

"Allah be praised!" He screamed back and jammed his hand in the pocket .

"Not today!" Meira declared, staring into the man's eyes. The sharp report of the Walther PPK followed, twice in quick succession. Both 9mm bullets blasted up through the man's jaw and brain and burst through the top of his skull, splattering blood, brains, and bits of bone across the wall. He slumped.

Meira's eyes were opened wide, staring down at the

motionless man, blood oozing from the gaping holes at the top of his head. She leaned back and slowly shook her head back and forth. Naseer Yassin, born 26 years ago in Hebron on the West Bank, a graduate of Cairo University, and a member of Hamas completely committed to their cause, was dead.

The whole confrontation took less than ten seconds. Finally realizing what was happening, Raphael bolted to Meira's side, his Beretta drawn. She was on her knees in front of the dead Yassin. Blood puddled under his body, spreading out on the pavement. Meira reached into the man's pocket and found what he was trying to retrieve.

"Here, Rafi, take it, a remote detonator. The bomb is in the leather bag. Don't touch it, the latches might be armed." Her lips quivered and her hand was trembling badly. All color had washed from her face.

He took the detonator in his left hand, his cell phone in his right. "I'm calling the police and the Jerusalem defense center. They should be here in minutes."

Sirens began wailing in the distance. Every police and military vehicle within 20 blocks was now speeding to Shammai Street. Raphael walked to the center of the street and yelled out. "The terrorist is dead, but the bomb is still here! Please, get up and leave the area immediately! Walk, please! Walk, don't run! Help each other! Please go to the sidewalks, leave room in the street for the vehicles!"

He returned to Meira. She was bent over Yassin's body, still on her knees, the pistol still in her right hand. She was shaking dreadfully. He reached over and eased the gun from her hand, pushed the lever down to engage the safety, and helped her stand. She waivered on her feet and he drew her to him, put her right arm on his shoulder, and with his left pushed the Walther back into her belt holster and snapped the retainer. An armored vehicle screeched to a stop within several feet of them. Three men jumped out in light blue shirts and dark blue tactical vests, weapons at the ready. Raphael guided Meira in their direction.

"Stay where you are! Who are you?" An officer barked, the muzzle of the automatic rifle pointed at Raphael.

"National Security agents."

"Mos... "

"Shhh... you know better, brother. I'm Raphael Mahler. She is Meira Dantzig." Meira weakly raised her head to stare at the officer.

"Oh... okay. Is she hurt?"

"No. But here, take this" Raphael extended his left hand. "It's the remote detonator. The dead man had it in his right jacket pocket. See the big leather bag against the wall? Be careful." He gestured toward the body.

"Did you shoot him?" the officer asked.

"No, she did. We were sitting here at the Babette Café having waffles and coffee. Meira noticed him first."

"The woman shot him?" He looked around at the all the people walking away from Shammai Street, then shook his head. "Ms. Dantzig, this would have been a huge disaster. You saved the lives of probably hundreds of people. Thank you so much for your courage. Ahhh..." The officer stopped to reflect, his left hand on his chin. "Any chance you two might have been the target?"

Raphael looked at Meira. "I hadn't thought about that yet. But no, I doubt it. My guess is it was random... they usually are."

"I agree," said Meira weakly.

"Okay, well, are you sure you don't need any help? I can have a car take Ms. Dantzig to the hospital."

"No, I think she's okay, just shaken up. Look, we're going back to Tel Aviv. File a report. I'll take care of her."

"Then go with God. And thank you again. The department

will contact you later. Shalom." The officers walked toward the body as another truck pulled up. Men with heavy woven coveralls stepped out, metal shields in front of them. The bomb squad.

"Shalom." Raphael turned to the street, easing the now nauseous Meira to the walkway. A dozen or more police and military personnel were now milling about the scene. The civilians had all but disappeared. Meira's hands were still trembling, her legs unsure. Her face was pale. He bent over and put her left arm around his neck. They stepped onto the walkway.

"Get me to the wall, Rafi. Please."

Meira dropped to her knees and doubled over on all fours. Her chest suddenly heaved. She retched, emptying her stomach. Raphael placed his hand lightly on her shoulder.

"Gees, I already paid for that waffle."

She coughed, then spit and barked a mirthless laugh. "Yes, we have a comedian!" She laughed again and coughed, wiping her mouth on her sleeve. She flipped around to the brick wall and sat against it.

"I killed him. I've never actually killed anyone before."

"Yes, Meira. He won't be making or delivering any more bombs. Like the police officer said, you saved hundreds of lives." He sat down on his knees next to her. "You just acted before your adrenalin kicked in. Now, it's catching up to you. Getting sick is a normal reaction. And girl, you are fast, I mean, really fast. I don't think I've ever seen anyone move that fast, take someone down like that."

"Come on."

"No, really. Remind me never to piss you off." He smiled.

"Hmmm." Meira turned toward him, leaned over and rested her head on his chest. She could feel her muscles relaxing. "Thanks, Rafi," she whispered. She snuggled her body against his and nuzzled her face into his chest.

"Meira?"

"Just give me a few minutes. Oh, this feels good. My stomach is getting better. Gees, you know, I think I could drift off. I'm dead."

"That's okay, it's normal. All that adrenalin in you is now draining away. But look, you can't sleep here on the street." His right hand brushed her hair back from her face. "Meira, I am so proud of you. I'll talk to General Goren and the director tomorrow. You were so fast, so precise. Just seconds. It was already over before I reached you." He paused, looking down at her. "But look, we have a lot of work to do at the Center. And we're still on the street." A look of helplessness filled his face.

"Rafi, please, just a few moments." Her head rubbed against his chest, her right arm inched up over his shoulder. "Please, let me gather myself. A couple minutes. Tomorrow is tomorrow. Today is today. Just a few..." She closed her eyes.

CHAPTER 8

Chugach Mountains, Alaska

A sliver of light teased through the opening of the tent flap. McClure blinked his eyes several times, attempting to adjust. The ever present early morning dreams of times past with Anne and Michael gently evaporated. He looked down at this watch… 7:30 AM. He'd slept through the whole night. Both his body and his mind had been so drained the night before that he drifted to sleep in just minutes. It was a good sign for his first morning in the wilderness. A light, cold breeze brought the scent of water. The Copper River was just ahead.

A clamor of high-pitched chirps assailed his ears. Dave unzipped the tent flap a little more and peered out. The owl was gone, but at least ten black-capped chickadees were hopping around in the middle branches of one birch, and some clay-colored sparrows in another. On the ground below four white and brown ptarmigans pecked at the grasses poking up through the snow in their daily morning ritual of foraging. He inched his body forward in the sleeping bag and poked his head through the flap opening. The sky was transitioning from black to gray. The air was wintry. He eased out of the bag and pulled on his boots, not an easy task in the cramped quarters. Crawling out of the tent, he slipped on his parka and zipped it.

McClure bent to his knees and glanced down into his fire pit.

A few embers still smoldered at the bottom. He quickly gathered some dead grasses and bushy branches and shoved them down and around the glowing embers, blowing gently until it took. The flames rose up the kindling and he grabbed a couple nearby branches from last night. More, heavier branches were positioned on top of the kindling. He had another fire going in short order. Now, what to eat?

McClure stuck his head back in the tent and rummaged through the pack. His face lit up. *Oh yeah, Mountain House® scrambled eggs and bacon… just add water, then a pinch of salt and pepper!* Cooking didn't take but a couple minutes. He sat back on the spruce boughs and slowly ate the meal, watching the sky brighten to the east. Gulps of hot coffee with a smidgen of sugar washed down his meal. The ptarmigans had quickly departed when the fire began smoking, but the chickadees and sparrows still hopped around the branches, checking out this curious new creature in their forest. He wished Anne could experience this. He felt there was at least a slim chance she would have passed on the resort hotel and slept in the forest with him. In a larger, more comfortable tent, of course.

His watch now read 8:30 AM. Once again he was lucky with the weather. Though this time of year the region was usually cloudy and foggy, especially in the morning, the sun was rising above the horizon in a blue sky. He doused the fire with the last of the coffee and pushed the walls of snow in and over it. Rolling up both the tent and sleeping bag was accomplished in short order. McClure slowly paced his way around the birches, retrieving the coiled nylon rope as he went. The stake sat securely in the notch of the trunk as he'd left it last night. Except for the cut spruce boughs and the now chilled remains of the fire, he was leaving the clearing as he'd found it.

He pulled the .44 magnum revolver from his holster, opened the cylinder, and checked the barrel. He re-holstered it, then reached for the Weatherby rifle and laid it across his legs. The barrel was clear, a round still chambered in the breech, and the rifle on safe. On went the snowshoes, the pack on his back and the rifle slung over his shoulder. He telescoped the hiking sticks, pulled out the compass and took his bearings. North and east… that's where he'd head. Find the river and cross it, then enter the most exciting region of his journey, the deep wilderness of Wrangell–St. Elias National Park. His path would stay well to the north of the Bering Glacier and the Bagley Ice Field. He pulled down his polarized goggles.

He started off slowly, breathing easy and pacing himself. Placing one snowshoe down after another. Swinging and planting one stick in the snow ahead of him after the other. He was on his way. The sun was above the horizon now and the snow reflected it intensely. It was hard not to smile at the gorgeous scenery. He felt the surrounding forest was welcoming him to its sanctuary. All of it was well beyond his expectations. As he went, the trail opened up again and he picked up the pace. Maintaining his rhythm and breathing didn't seem as arduous as it was yesterday. He regularly swept his glance to both sides, scanning for animal sign. He could tell he was progressing well and he felt at one with wilderness.

In less than an hour, he emerged from the tree line onto the banks of the Copper River. Not a wide river at all. Several narrow, pebbly sand bars jutted out from the banks and peeked occasionally through a layer of snow and ice. The water was extremely low and would be no trouble at all to cross. And, there was enough of the snow that he didn't have to remove his snowshoes. He carefully picked his way up a pebbled bar to where it sank into a shallow stream of shimmering water. Squatting down, he laid down his right-hand stick, took off a glove and cupped it into the water for a sip. Ice cold and delicious. He bent down and scooped up another. His thirst slaked, he stepped into the stream and delicately crossed the ten foot expanse with no difficulty. In minutes, he strode up the bank and entered Wrangell–St. Elias, a wilderness area deeper and more rugged than most in North America.

Another two hours of fairly level terrain brought him to a rising meadow of sorts, a huge section of grassland covered with snow. Toward the south, he saw where the trees gave way to open space that appeared to stretch on forever… the Bering Glacier, extending down to the Gulf of Alaska. McClure knew the Bagley Ice Field was somewhere adjacent to the glacier and a tad northeast. Across the shimmering expanse in front of him, Dave could see the forest picked up again, so he turned back to a spot just inside the cover of the woods behind him. He dropped his pack to the snow and laid the rifle against a near birch. Lunch. No fire this time, just some beef jerky, a piece of chocolate and a short rest. The woods smelled wonderful. He sat on the pack and ate, considering. The morning had gone very well. He figured he had about three hours of strong daylight left. He swigged down the last of the water in his bottle, then stuffed snow in it to create more.

Fitted once again with his pack, rifle and sticks, McClure plodded across the field of snow to the opposing woods. The sun on the snow was dazzling, an intense sheet of light. Thank goodness for the polarized lenses. The field was larger than he thought and took about forty-five minutes to cross. When he entered the woods, it felt good to be back in the forest. He'd felt exposed out on the meadow. He didn't understand why he'd felt that way, so vulnerable to whatever, the sensation was just there. He'd made a lot of progress on this second day, and planned to do maybe another two hours or so of snowshoeing before finding a campsite. Looking skyward, he noticed the sky was clouding over and the temperature was falling in tandem.

McClure paused. Directly ahead through the trees he

could make out large patches of bright blue and orange. The sun flashed off the shiny surfaces. Snowmobiles. Two parka-clad figures stood silently by them, watching him. Their broad shoulders told him they were men. What the hell? Dammit, he'd been distracted, not paying enough attention to the woods ahead. Squinting his eyes against the glare, he slowly slid the Weatherby off his shoulder.

"You can stop right there, mister. White men with guns make even friendly Indians like us nervous. You move that muzzle toward us and you'll be dead before your finger reaches the trigger."

"Who are you?" McClure barked, re-shouldering the rifle.

"You can come closer." The larger man on the right spoke. He was sitting on the fender of the orange snowmobile. McClure could see a rifle held low across the man's thighs. The other man, a lot younger, held a rifle too.

McClure tentatively stepped toward them, scanning the vicinity as he approached. He didn't see any others. He was ready to drop down to the snow if the man swung his rifle toward him. He was now about ten yards away.

"Closer please."

A few more steps and they were now close enough to reach out and touch each other. McClure stopped. "Who are you?"

"I'm Robert Ewan. This is my son, John. And you are?"

"I'm Dave McClure. What do you want?" McClure looked the father over. His face was weathered, and his salt-and-pepper hair was tied back in a ponytail. He wore what looked to be a quartz beaded necklace. Other than that, nothing else indicated his Indian heritage. His parka was a silver Columbia Sportswear brand. His son looked like any other young man, probably in his teens.

"Well, we've just been sitting here, curious as hell to see what creature is making all the noise and scaring away the game for miles. We've been hearing you since you crossed the Copper River this morning. We lost a moose we were tracking. So, we came down south a bit to see what kind of beast you were," Robert said drily.

"I was that noisy?"

"Noisy enough that we thought maybe you were practicing to march in the Macy's Thanksgiving Day parade. Oh yeah, we Indians have TVs now. I like to watch that parade every year on my 46" HD Samsung."

"Gees, I'm sorry. I haven't been in the woods or snowshoeing for a while. But you know, both of you could go to prison for twenty years for threatening a federal officer, special agent. I've been away for a while but the law still applies."

"Federal officer, special agent? Well, shit, we didn't know we'd run into the Lone Ranger. That means Tonto must be around here somewhere." He paused, hunched forward as though peering out into the woods. "No, I don't see him. I guess we'll leave your scalp alone. From the looks of it, Dave, it really wouldn't look that great on my lodge pole anyway. We don't want any part of a special agent. So, I'm sorry too. We're just cautious around white men. Even now."

"What kind of Indians are you, Robert, what tribe?"

"Ahtna. You probably never heard of us. The Ahtna Tribe is part of the Athabaska Nation. Ahtna means 'ice people' in Athabaskan. We spend a lot of time here south of the Copper River, near the ice fields, and a little north too. I'm teaching my son to hunt. He already got one moose and we were going after another, that is, until you scared everything away. See the packs on the back of my son's snowmobile? That's the meat. We'll give you some if you like."

"Yes, thanks, I would. I've never tasted moose. Okay, well then, I'm just heading east for some camping."

"There goes the neighborhood. Hey Dave, you seen any

other terrorists?"

"Terrorists? What do...?"

"Dave, it's like this... my tribe, my nation, we've been fighting terrorists for 200 years. First, the Russians. Then the Americans, like you. We lost. Now, you tell us where we can live, where we can hunt, and where we can catch fish. We used to be able to figure that out pretty well for ourselves. That's just a social commentary I make whenever the opportunity comes along."

"I understand, Robert. I don't understand that myself. I don't support that situation either. Ah, by any chance would you know Jack Stroud?"

Robert glanced over to John. "Yes, we know Jack Stroud very well. He's out here often. Stroud is a good man. I don't know how well you know him, but Jack lost his wife to cancer about five years ago. He never did get over it. I think he comes out here in the wilderness to be alone, find some quiet, maybe find some peace. He respects the wilderness and the creatures in it. He leaves no mark of his passing through the forest. In that way, he's kind'a like us. How do you know Jack, Mr. McClure?"

"Ed Weiss told me to be on the lookout for him. Stroud is Ed Weiss' brother-in-law, but then you must know that. So, have you seen him?"

"No, and that's strange. I had no sense of his presence. I know that must sound weird to you, but I have that gift."

"From the Great Spirit?"

"Touché, Dave," Robert chuckled.

"Anybody else out here now that you're aware of?"

"Not that we've seen. I thought I smelled smoke yesterday, but it was very faint. Maybe I was just imagining things. We've been skimming north along the edge of the Wrangell-St. Elias, so I can't say what's going on south of here." Robert turned to his son. "John, give Mr. McClure one of those smaller packs, the tenderloin. Dave, you'll like that. It's delicious."

"Thanks very much, Robert. I can't wait to grill it."

"You're welcome. I think you're a pretty good white man too, Dave. Sorry about that Lone Ranger stuff. Look, don't hang this meat in a tree. A couple of male grizzlies out here wander around even in winter. They'll smell it and come tromping into your camp looking for it, and probably eat you instead. Keep what's left wrapped and bury it a foot under the tundra about ten yards from your campsite, and then pack a lot of fresh snow over it. That will conceal any smell. By the way, snow's coming. Maybe a couple days."

"Really? How do you know that?"

"Another one of those gifts from the Great Spirit, Dave." He smiled. "Out here, a snowstorm can be a serious thing. Watch what you're doing and try to make sure you're not caught out in it. I have a feeling it's gonna be the first big one of the season."

"Well, thanks again, Robert. You take care also. I'm pleased to have met you." They shook hands. He watched as they climbed on their snowmobiles and started their engines. McClure was astonished at how muted they were.

"Say, how come those snowmobiles are so quiet?"

"These are Arctic Cats that I had customized right after I bought them, Dave. They have an acoustic blanket shielding them, open cell foam and a heat resistant fabric. That's topped off with a super quiet muffler system. They're great for hunting. We can move through the forest like whispers in a wind. We Indians have a lot of modern conveniences now." He laughed. "Say, if we take another moose, we may come by your campsite and drop off another cut of meat for you."

"Sounds great. How will you find me?"

Robert laughed again. "I really don't think that's gonna be too great a challenge! You take care now, be careful, you hear?"

"You too, Robert, John. Take care." They waved and Dave

waved back as they turned their snowmobiles north, whispering their way through the trees. *Moose. Boy, I can't wait to try it. Thanks guys,* he thought.

McClure noticed that the tracks of the Ewans' snowmobiles came from the north and they had gone back in that same direction. He shrugged his shoulders and turned back to his easterly path.

CHAPTER 9

Wrangell-St. Elias Wilderness, Alaska

This wilderness was starkly different. After another hour of trudging through the snow, McClure still hadn't seen any animal sign, nothing since the ptarmigans this morning. No bears, moose, caribou, or wolves, nothing. And he suddenly realized that after seeing them pretty much everywhere all day, even the small birds now seemed absent as well. The woods had grown quiet, dead quiet except for the sound of his sticks, snowshoes and breathing. *Part of being in the wilderness,* he thought.

The forest had transitioned from the earlier mix to all black spruce now and the trees were spaced further apart. He saw a little clearing ahead and made for it, thinking it was about time to find a campsite. As he drew to within about twenty yards, he saw movement on the ground. Something reddish in color, between two and three feet long, appeared to be pulling at something in the snow with its teeth. Dave slipped his Nikon binoculars out and took a look. It was a red fox. Something dark brown was jutting out of the snow in front of where the fox was biting. It looked like a boot.

McClure warily approached the clearing. The fox's ears went up, it turned to face him and let out a warning yelp. Dave stopped. The fox swiveled its body, turning toward him, now growling and hissing. He reached under his parka and drew the revolver. Okay, yeah, a .44 magnum might be just a bit heavy to shoot a fox with. But if this little shit charged him, he was ready to blow his fluffy red ass into the next time zone. He inched forward, the two trekking sticks in his left hand and the revolver in his right. He was only ten yards from the fox which had buckled its legs and settled its front paws, ready to leap.

"You think you want to mess with me, you little shit!" McClure shouted at the top of his lungs and waved the two sticks above his head. That did it. The tiny humps of its eyebrows rose in alarm, and it was suddenly clear that the fox didn't want any part of McClure. It gave up its discovery and abruptly bolted off to his right at amazing speed, kicking up snow. It was soon out of sight. Dave turned his eyes back to the brown boot sticking out of the snow. He drew closer, approaching gradually, step by step. Kneeling down, he reached over and brushed away some of the thick layer of snow. A body.

It looked like a man's body face down in the snow. The fox had also apparently dug out some snow around the man's head and had chewed on his neck. McClure leaned his rifle against some low spruce boughs, then unsnapped and slipped off his pack. He continued to brush the snow off the body. The snow was heavy and clung to the body in clumps. Interestingly, the man's rifle, a Winchester, was lying next to him as was a backpack. Dave brushed everything clear. He estimated the man was about 5'10" and of fairly stocky build, wearing a light gray parka and a dark blue wool stretch cap. He guessed it was probably a hunter who suffered a sudden heart attack and maybe had just enough time to take off his pack before collapsing.

As he slowly rolled the corpse over, the first thing he noticed was that the man's eyes were wide open, frozen into the snow. The tissue of his eyelids was stiff, no way to close them. The vacant stare was surreal, and a shiver went down Dave's back. He pushed and brushed snow off the front of the parka and suddenly saw a dark stain across the dead man's chest. Looked like blood... what the hell? Dave's gaze drew inexorably up toward the head and he involuntarily sucked in a breath. The man's throat was slit from ear to ear. The cut was so deep that barely another inch would have severed the spinal cord, decapitating him. The only things holding the head on were a few bones and the cold that had fused everything together. Nope, no heart attack here.

Except for the horrific cut at the neck, the body was intact. Apparently, the only wildlife that had discovered the body so far was that red fox. The heavy cover of snow must have had something to do with it, cloaking any odors. And of course, the corpse was frozen solid, almost no decay at all. Dave began digging through the jacket pockets. He found a set of keys, some loose change, a box of twenty .270 Winchester rifle rounds, and a couple sticks of chewing gum.

The untouched rifle and ammo indicated that the attacker hadn't been after them. Dave reached around the corpse to the left pants pocket and found the wallet, then sat back in the snow and opened it. It still held $70, so again, this was obviously no robbery. Dave slipped the glove off his right hand and pulled out an Alaska driver's license issued to a... Jack Stroud. Damn. Ed Weiss's brother-in-law.

Well, Dave now faced two tasks when he got back to Valdez - a trip to the Sheriff's office to report an apparent homicide, plus an unpleasant notification to the Weiss family. He surveyed the body once again. Nothing else about the corpse or the immediate surrounding area struck him as abnormal or interesting in any way. The killer wasn't looking for money, nor for a rifle and ammo to sell to get money. Dave assumed that the killer just wanted to kill Jack Stroud. But why kill an old guy who lived alone and found solace by escaping now and then into the wilderness? Why? What was the point? Maybe Jack had run into some drunken hunters or had taken a shot at some moose they were tracking and pissed them off. Well, the 'why' was something the police would have to solve.

His eyes caught what looked like a gold chain around Stroud's neck, matted in the clotted blood below his throat. Dave gingerly pulled it up. A gold cross and a locket. His big, clumsy fingers were a little numb from the cold so it wasn't easy, but he finally managed to open the locket. A picture of a woman smiled at him from inside. Probably Jack's wife. He placed the cross and locket back below the dead man's shirt, then sat there on his knees for a moment looking down at Stroud. He shook his head in disgust. What a way to die. Dave put the wallet and key chain in his own pocket, and then retrieved the camp shovel strapped to his pack. He stepped off an area the size of Stroud's body and began thrusting the shovel into the snow and underlying tundra.

It would be a shallow grave only about two feet deep, and of course, temporary. Sheriff's deputies would have to come retrieve the corpse, and after a coroner's cursory review (the cause of death being obvious), re-bury it in Valdez. That is, unless Jack Stroud had a will that specified differently, but McClure doubted it. Digging the grave took him a little over an hour. He unstrapped the sleeping bag from the man's pack, unzipped it the full length and wrapped the body in it, making sure Stroud's head was covered, then dragged corpse, face up, into the rectangular hole.

Dave sniffed at the rifle's muzzle. The pungent odor told him that it had been fired, but no rounds were left in the magazine. He wrapped the rifle in the small tent he found in Stroud's pack and laid it next to the body. He fumbled inside the pack again and found some jerky and dried food packs, and stuffed those into his own pack before placing the backpack in the grave. Lastly, Dave shoveled dirt and snow back into the hole until a moderately raised mound marked the grave. In some ways, he'd just buried a kindred spirit, a man who had been whittled down by years of grief, a man who came here into the forest to get away from it.

McClure then did something he hadn't done in well over a year, not since the dual funerals of his wife and son. He prayed. He stood looking down at the grave and whispered: "Lord, this is your child, Jack Stroud. I don't know much about Jack, but I suspect he was a pretty okay guy. Anyway, Robert and John know him better and they say he's a good man. Jack came here into this glorious wilderness of your creation to find peace. Instead, he found death. Lord, please open the door for Jack and welcome him into your house. I ask this in your name, Lord. Amen."

There was no need for a cross to mark the grave. After all, it wasn't permanent. Instead, McClure circled inside the little clearing with his knife drawn, cutting off boughs and strips of bark on about six trees to mark the site. After taking a swig of water, he glanced at the sky and figured that he had about an hour and a half of solid daylight left. Once again, he scanned the clearing.

The indentations that had been filled in with snow looked like footprints. He could still make them out. They all came from and went back to the southeast. However, other than the small, more recent skittering tracks of the fox, these tracks in the snow had come between many of the trees from around the clearing. There were a lot of them. Dave was fairly certain that this murder was not the work of one man, but of several men. There were just too many indentations, too many tracks. He stepped out of the clearing to his left and about fifteen yards away, there was another set of tracks, parallel to Stroud's that finally turned and broke toward the clearing. So, they had stalked him, chased him, surrounded him, and then right here, had cut his throat, killing Jack Stroud. Bastards.

As he turned about, his eyes scanned 360 degrees through the forest. The killers might still be in the vicinity, and McClure didn't want to run into them. He would change his direction to the northeast in order to put some distance between them.

Then the thought struck him. *Whoa. Wait a minute. McClure, just what the hell are you doing?* Dave turned in place, glancing at his trail leading eventually back to the Copper River, the Richardson Highway and Valdez. *Shouldn't you be going back to Valdez to report a homicide? You are MR. McClure, you know, not Special Agent McClure, and you don't have a badge. Not anymore. You haven't had one in more than a year. What's going on with you? What are you doing?*

The neurons in McClure's brain began to spark and fire on different pathways. He hadn't felt that in a long time. Going back to Valdez would take over two days. And it'd take the sheriff and his deputies another two days to get out here. Well, no, he supposed they could fly out in a helicopter. but it would still be two or three days at least. The murderers might be well away from the wilderness by then. They might never be caught. *Is that what I want?*

No, no way. Dave was no stranger to violence. Ten years in the FBI had made him numb to that. No, he was going to track these sonsofbitches down and see to it that they face justice, either in a courtroom or out here in the wilderness. Okay, yeah, he was outnumbered, but he had the advantage of surprise. They, whoever they were, didn't know he was here. He had a .300 Weatherby and .44 magnum. Hell, he was loaded for bear. These assholes were going to pay for what they did to Jack Stroud.

His snowshoes and pack strapped on, McClure slung the rifle over his shoulder and turned to the north. Somewhere out there he'd make another campsite. No fire tonight. The moose tenderloin would have to wait. He'd use the small Sterno pack stove to heat some chicken soup and coffee. Didn't want to risk sending smoke into the night air that might drift down toward the bay. No perimeter alarm tonight either. Dave was willing to risk a chance encounter with some wandering critter rather than allow a clatter from the rope alarm reverberate down the valley and alert the killers to his presence. No, he needed to ensure that the element of surprise would stay on his side. These guys were about to face a different brand of violence... Dave McClure's brand. Indeed, these murderers were going to have a 'come-to-Jesus' meeting. Assholes. Whoever they were.

CHAPTER 10

Amsterdam, Netherlands

He sighed, peering out the window as the KLM plane nosed up to the gate. Time to get up. The flight from Moscow to Schiphol Airport on the southwest edge of Amsterdam was on time. It had been a little over three hours, not too bad. Viktor Nechayev swallowed the last of his vodka over ice, handed the glass to the waiting stewardess, and unsnapped his seat belt. As the forward hatch door opened, everyone in first class rose. He slipped on his suit jacket and proceeded to the exit as the stewardess smiled, handing him his overcoat.

"*Specibo.*" 'Thank you,' he grunted as he bent slightly through the hatch door.

He wound his way to the airport entrance, putting on his tan camel hair overcoat as he walked. He skipped the baggage claim, steering directly to the glass and steel doors. Somebody would get his suitcase. It's hard to miss an elegant, steel-gray, multi-wheel Rimowa suitcase. The Germans know how to make luggage. As he opened the glass door, he noticed three men immediately to his left dressed in black jackets, blue berets and matching pants, all toting submachine guns. Must be an elevated threat level, which happened occasionally at Schiphol. He glanced at his watch - 11:10 AM. Two men approached him directly, nodding.

"Privyet, Gospodin Nechayev!" 'Hello, Mr. Nechayev!' The man in front said cheerfully.

"Privyet!" Viktor replied.

"The cars are right over here, sir." Nechayev turned to the man's gesturing hands, pushing up his overcoat's collar as he went. The brisk wind quickly disheveled his hair. The men flanked him, their coats remaining open, flapping in the wind in the unlikely event that they might have to draw their weapons.

"Windy. Very brisk. Nothing I'm not used to, though." Nechayev exclaimed, thinking of his apartment in Moscow. There were more than enough blustery days there too.

"Yes sir, it will be this way all winter, winds up to 50 mph, plus the cold," the lead man replied.

Three silver Mercedes S Class Turbo sedans sat in line, purring at the curb. One man stood at the front fender of each vehicle. They nodded respectfully to Nechayev. The lead man quickly stepped ahead and drew open the rear door of the middle Mercedes.

"Pozhaluista." 'Please.' The man gestured to Nachayev.

"Specibo." Viktor climbed onto the plush leather seat. His door shut softly, then he heard the other car doors slam, and the three Mercedes swung smoothly away from the curb in tandem. Viktor combed his hair with his fingers. He was 55 years old, a big man, 6'3" tall, with large shoulders that implied great strength despite his distinguished looking apparel. The three vehicle mini-motorcade soon swung onto Highway A4 heading generally northeast into the heart of Amsterdam.

"How long?"

"Short, only about twenty minutes to the harbor front, sir."

"Great. Thank you. Anything that I can do for you guys?" Nechayev asked.

The man in the front passenger seat responded. "No sir, actually, the embassy treats us very well. This is great duty. But thanks very much for asking. The prime minister himself sent a message to the embassy saying we should strive to meet all your needs. It is truly an honor to meet you, Mr. Nechayev." Nechayev was well wired politically.

"Well, this is first class treatment, men. Look, if any of you ever wants another job, and a better one at that I might add, give me a ring."

"Yes sir! Thank you for the offer!" The driver and the shooter both said heartily, broad smiles on their faces.

Viktor had the ability to use grace and levity even on a serious mission. Years of having to respond cheerfully in situations while at the same time being prepared to use his weapon without warning had steeled his personality. Today's meeting would be more of a confirmation, a validation of facts as they'd already been related to him. Viktor considered Leonide Krasnoff a good man, just perhaps a little too Europeanized. And of course, that short coming could be fixed.

"Krasnoff?"

"He's waiting for you, sir."

"Excellent again. Good work, men."

The lead man in front turned to speak directly to Nechayev. "Sir, we will be merging onto Highway S114 and then to Iburglaan Street, which will take us down into the Piet Heintunnel under the harbor. The tunnel takes only a few minutes. It's completely safe. Then we'll swing over to the Harbor Club on Cruquiusweg. Nice place. It's right on the water, an extension of the harbor area."

"Good, thank you. I look forward to it." But he wasn't smiling. Viktor didn't like tunnels. It was an old fear, instilled in him by his father's stories of hiding from the Nazis during World War II. In the forests, his family and their neighbors dug out and hid in earthen tunnels that frequently collapsed, burying many people alive. Over time, the boy had developed a fear of tunnels that dogged him all his life. But Nechayev was willing to put up with them occasionally if it was just for a few minutes. Nonetheless, the dark cavern of the tunnel made him shiver.

The three silver Mercedes emerged from the tunnel, and Viktor sighed with relief. They veered to the left, then took a left again onto Zeeburgerkade which in time became Cruquiusweg, paralleling the harbor as they zipped along. It was a beautiful sight, well worth coming to Amsterdam just to see it. Viktor noticed empty white chairs and matching tables facing the water behind a white fence outside the Harbor Club as they pulled in front. It was too cold and gusty for anyone to be sitting out there now. Two men jumped from the front and rear Mercedes, one stopping at the front door and one going inside. Minutes later, the man, a shooter, came out and gave a thumbs up.

Krasnoff was seated in a curved booth near the rear of the main dining room. A few booths were already occupied by people out for an early lunch, but the place was still mostly empty. The lunch hour wasn't yet in full swing. Nechayev smiled as he walked toward Leonide, admiring all the vivid colors of the nouveau art decorating the walls. The Harbor Club was indeed a very nice place.

"*Privyet,* Viktor! Good to see you!" Krasnoff stood to greet him.

"Good to see you too, Leonide. It's been too long."

They settled into opposite sides of the booth. A bottle of vodka chilled in a bucket on the table along with two glasses filled with ice. Krasnoff was alone. That was a little unusual, Nechayev thought, but then Amsterdam wasn't too risky a place. The city government and their police force saw to that. Too many tourists and visitors came to the city and also to the Hague not to. And, the Harbor Club was classy, so he probably felt comfortable here. Viktor himself never traveled alone, no matter how safe a place was touted to be. But then, he had a few years and several close calls on Leonide.

"Safe?" Nechayev asked.

"Yes, I have this place scanned discreetly by the embassy's tech people every month or so. We're good. It's a little cold and windy outside for your visit today, Viktor, but I thought you'd like the place."

"I do. Good choice. Cold and windy? Come on, Leonide, you are doing a great job, of course, but you've been away from Moscow too long, away from Moscow and the Russian winters. Right now, you could subtract 20 degrees and add 20 mph to the winds in Moscow. Perhaps you should come back for a while, eh?"

"Oh, I remember Moscow and all the snow and wind very well. I was just speaking in relative terms. Besides, I love the Indonesian food here. It's delicious. We can order some, if you like. And, would you like me to pour some vodka?" He gestured to the bucket.

"Well, maybe we'll let you stay here just a bit longer. But... no thanks to the vodka. I had some on the plane. And my stomach couldn't handle spicy food right now. Thanks anyway."

"Are you staying over?"

"Yes, the Park Plaza Hotel. I want to get some rest after this. I've got an early flight back in the morning."

"All right. Well, how would you like to begin?"

"First thing, the money... $200 million all together, yes? We received the down payment. Very good work, Leonide."

"Viktor, you should have seen Madani. He was beside himself. He and his government can't wait to get their hands on the weapon. He assured me that the final payment will be wired to the Zurich account within 72 hours of delivery."

"Great. That's what I came to hear. And rest assured, we will wire your payment of the deposit, $25 million, to whatever account you identify within 48 hours of our receipt of the final payment. Okay?"

"I can't thank you enough, Viktor."

"You've done a great service for us. This is a big deal. The money will allow us to increase our operations in several countries and absorb some competitors or move them out. A new offensive. Are you sure you want to retire? Does Vassily still plan to leave the military and go with you?"

"Yes. He's disappointed about not receiving a promotion that he expected. He had all the schooling, the training… he was on the right track. Anyway, he's going to resign his commission and tell them that he wants to try something else."

"Okay. Well, where to?"

"We're still looking. Maybe Thailand, down near the ocean resort areas. Or, the Philippines. The mountains there have some nice towns… like Baguio. We can live very well in those places."

"The Israelis will come after you with a passion, no matter how tight we try to keep the information. You know that."

"Yes, I know, I've thought about that. But Viktor, I truly don't think the Israelis will be much of a problem to anyone if the Iranians launch successfully. And they're certain that they will. There will be no more Israel, no Palestine, no Mossad, not for years anyway. We'll be okay. We'll have the money to put some good security in place."

"I'm happy to hear your confidence. I will miss you though. You've done so much for the organization. Now, are you absolutely certain that this will appear to be purely an accident. No signs of combat wounds on the security force personnel?"

"Vassily says he guarantees it. The Americans have been having problems with their nuclear surety program, it's no secret. The news media made sure the whole world knows that. His team will be using EMP rifles, Electromagnetic Pulse weapons. They got the original technology from the Americans in the mid-1990s. He says he can't divulge any specifics, but the pulse weapons have been improved to where they are strong enough to interfere with a person's atrial and ventricular conduction."

Puzzled, Nechayev tilted his head and furrowed his brow.

"Basically, sharpshooters make two or three shots to the skin of the target, and those EMP pulses interfere with neural activity, eyesight, speech, and more importantly, stop blood flow between the atrium and the ventricle. The target dies in seconds from heart failure. They wear large battery packs on their back and waists to support the charge."

"No marks on the bodies?"

"None. The bodies will be placed back in the train car, and it will appear that they died in the crash. Vassily's men will ensure that the locomotive's diesel fuel tanks are ruptured. The train will burn to a crisp along with the security force."

Incredulous, Viktor's mouth gaped, startled by the prospect. "Leonide, such rifles, unbelievable! Is it possible to obtain a few, just a few? They would be an extremely effective tool for removing competitor leadership... imagine the possibilities. No evidence."

"I already asked. Vassily said that would be impossible, at least for now. Not under his command."

"Too bad. Ah, nuclear contamination?"

"Absolutely. The containers will melt. It will be a terrible, radioactive mess. And the pressure on the American president to eliminate their nuclear forces will be immense. This of course is the official, core rationale for the mission. A psychological operation using the train crash and contamination of their beautiful Alaskan wilderness to squeeze the president and American Congress to act resolutely and minimize future risk. And for us, securing the weapon and delivering it is key. The plan should work flawlessly. I have my brother's word."

"And the Americans, they won't notice the missing

warhead?"

"Not for months, if ever. The area will be a disaster. As I said, the locomotive itself and the cars will burn to a crisp. The security force will be nothing but burnt bones. The nuclear containers will all be ruptured and scattered. And, if eventually they find something is awry, they will keep it extremely close, highly classified. They'll be both clueless and helpless."

"How soon?"

"The team should already be in place. They were transported to the site about four days ago. They're canvassing the area, and should be ready to launch the operation soon. Vassily was reluctant to give me the exact time frame. I trust him. If he says they'll be on schedule, they'll be on schedule."

"Leonide, this is all excellent. It's what I was hoping to hear. Our government will of course act shocked and dismayed by the events and then will pressure the Americans along with the rest of the world to clean up their nuclear surety program and reduce their nuclear force. But among themselves, they will not be unhappy. An understatement, I'm sure. No pressure will be put on the organization at all. And as I said, this money will allow us to expand and do so rapidly. Good all around."

"A toast?"

"Okay, but just one." Nachayev laughed and Krasnoff joined in. Krasnoff poured and they raised their glasses to their imminent success.

"*Vashe zdorovie!*" 'To our health!' Broad smiles washed across their faces as they clinked their glasses and drank.

CHAPTER 11

Wrangell–St.Elias Wilderness, Alaska

McClure found it quite a bit harder snowshoeing the uphill trek to the north-northeast. The incline sucked the energy out of him. The woods changed again too. What had been since yesterday mostly well-spaced black spruce were now thick stands of birch and aspen with no trails at all, so it was pretty tight going. The forest was growing darker by the minute, but he had to put some distance between himself and Stroud's grave. And of course, between himself and Stroud's killers, whoever and wherever they might be, most likely someplace to the southeast.

He adapted to the constant rise and his breathing shifted accordingly, taking on a slower but steady pace. He looked at this watch… after three o'clock. He'd give it another hour before stopping to make camp by the pale, ambient light of the early evening deep in the winter forest. He saw an occasional rabbit track in the snow, so there was indeed some wildlife in this area for snaring and cooking. A wandering thought struck him, what would Anne think of what he was now doing? No doubt about it, she'd say he was out of his mind. Maybe he was. But for the first time in over a year, Dave had a reason for continuing, a purpose in life. Jack Stroud's murderers needed to be found and brought to justice, and McClure was the only one at the moment who might have a credible chance in doing that.

By four o'clock, darkness was closing in on the forest. McClure noticed the ground was leveling off. He stopped in a small glade that looked just big enough for his one-man tent with room to squat in front. The area was all birch and aspen now though, so he'd have no nice spruce boughs for a mattress. He'd have to find some bushy twigs and brush instead. His eyes swept the area and it was apparent those were readily available. Dead ground cover poked above the snow everywhere he looked. He laid down the softest brush he could find for a makeshift mattress under the tent. As night fell, he found that his eyes grew accustomed to the darkness. The tent went up in about five minutes with the sleeping bag stretched out inside. In also went the rifle and the backpack. He took the pack shovel, a plastic trash bag and the moose meat and stepped off ten yards beyond the camp's perimeter. He dug out about a foot and a half of tundra, put the packed meat in the trash bag and buried it like Robert had suggested. Lots of fresh snow he heaped on top and marked the nearest tree with his knife.

Dave pulled the little steel Sterno kit out of his pack, set it up and lit it. It took immediately. The chemical had a noticeable odor, but slight enough to ignore. A fire would have smelled much stronger, perhaps perceptible even a couple miles away. The glow from the little Sterno was actually inviting, somewhat calming here deep in the woods. He poured some water from his bottle into a metal cup and put it on the stove for coffee. He'd have some chicken soup after that. The forest was quiet, no sound at all except for a slight rustling, a clicking of branches in the breeze. The sky was cloudy but a faint radiance near the horizon showed where the moon was rising, so the clouds weren't too thick.

Tonight, his second night in the wilderness, McClure no longer felt like an intruder. It felt good to be sitting here sipping coffee, watching the trees and the snow under them. He was comfortable with it. The soup was beginning to swirl and bubble in the little pot. The aroma made his mouth water. He scooted himself around and reached back into the tent for his pack. He found a Chips Ahoy package in an outer flap and returned to his spot outside the tent. No law against having dessert before dinner, was there? Cookies and coffee were meant for each other, no doubt about it.

As he crunched down on a chocolate chip cookie, Dave realized that he no longer thought of Anne and Michael as just dead and gone. Now, he was remembering only good thoughts, almost as if they were right here around the fire with him. The pain of crippling grief had eased away, no longer gripping his psyche or tightening his stomach. *So what has changed?* he began to question. He loved Anne and Michael both as much now as he ever did. He still missed them, true.

This trek into God's wonder of creation, the Alaskan wilderness, was shifting his perspective. He hadn't recognized it until now, but it was working. Well, maybe Pete Novak was right about this… that on his own, he would eventually find a way to live with his grief. Could that be what was happening? Was coming here into the Chugach Mountains and the Wrangell–St. Elias purposeful after all? The thought of dying here deep in the wilderness, even the possibility of it, had not crossed his mind since he'd left Valdez.

David looked up through the branches to the sky above and whispered, "You do work in mysterious ways, don't You?" His lips creased in a smile.

The last cookie went down with the last gulp of coffee. And the soup was ready. Great timing. He literally gobbled the soup, hardly breathing as he shoveled in spoonful after spoonful. He was completely invigorated by the meal and by a new acknowledgement that he really was beginning to feel better about life itself, his life.

He was so wide awake, sleep seemed impossible. The last thing Dave wanted right now was to crawl inside the tent, squeeze into the bag, and try to sleep. The temperature was dropping, probably now into the teens, but that didn't bother him. He loved sitting here by the glow of the little steel stove, just thinking. Then, creeping out of a dark corner of his mind, the image of Jack Stroud's body interrupted his communion with the wilderness. It was a sudden and sobering thought.

Dave turned to those solemn considerations. Just what exactly was he going to try to do when he found the murderers? He didn't know how many of them were out there. They killed Jack Stroud with a knife, but out here in the wilds of Alaska they'd have to be brain dead not to have rifles. Yet he hadn't heard any rifle shots since he'd stepped off the Richardson Highway and onto the trail. Game had to be abundant out here, but he hadn't heard a single shot in two days. So what then? Had they left? He doubted that. And who the hell were they? Escaped convicts? Hardly. Ed Weiss would have mentioned that for sure.

The only thing Dave really felt certain of was that there more than one. That was it. How many was impossible to figure at this point. They had to be camping with tents, or else have a cabin. Stroud was at least two to three days dead when he found him, and it took McClure two days of snowshoeing to do that. So, Stroud was killed three days ago? Maybe four? Again, impossible to figure since the corpse was frozen solid. This was really all he could speculate about now.

Despite all the uncertainty, McClure knew he still had a huge advantage. Surprise. Dave knew the bad guys were out here somewhere, but they had no clue that he was. No one except Ed Weiss and the two Athabaskan Indians knew he was out here. So tomorrow, he would begin a search of the area to the south and southeast. He would have to move carefully, silently, and that meant traveling light. It had seemed apparent to him when he scanned the area that Stroud had tried to run, but unsuccessfully. Dave hadn't yet seen from where Stroud had run, but the killers outran him, cornered him and killed him.

In case he had to run, Dave would take only his weapons, water, compass and binoculars. Maybe some beef jerky and crackers. The pack and the rest of his stuff would have to stay here. He looked around assessing the campsite. It would work as his base camp. This was where he'd return each night.

First thing in the morning, he'd travel further east and south, then after a while, come back slightly toward the west. A wide loop. That would give him a feel for the surrounding territory, and could be an advantage too… finding some spots to set traps or deadfalls. If the killers were south or southeast of him, they would never expect any human coming toward them from the west. Game maybe, but not a man. There was nothing further due west but treacherous ice fields and an exceedingly rugged, rocky wilderness.

He had the Weatherby .300 magnum rifle with a Leupold scope which, as Ed Weiss commented, would allow extremely accurate shots out to 800, 900, even 1,000 yards. He also had a powerful .44 Magnum revolver with platinum-tipped hollow points for closer confrontations, if it came to that. And, he had plenty of ammo for both.

McClure also knew he was very fit, more so than most men. He was fast with his hands and could still prevail in most close-quarter combat situations that he could imagine. He felt sure that, combined with surprise, he could wreak some serious havoc on these bastards, gain the upper hand if he worked it right. However, if there were more than two or three of them, he'd have to be more innovative. But then again, why would that many men be out this far, anyway? They're not shooting, so they can't be hunting. What then?

Well, that particular 'what' didn't really matter right now. What mattered was that he locate these men using as much stealth as possible, and figure out how to overcome them. Bring as many of them back to civilization and justice as possible. That is, if any of this was at all possible. He'd give it one hell of a shot, at any rate.

Dave cleaned up his utensils, packed everything away, and covered up the area with heaps of fresh snow. After a last look around the area, he crawled in the tent, took off his parka and boots, and crept into the sleeping bag. He zipped the front flap half way down, allowing a sliver of moon light to trickle in. He thought of Anne and Michael. In his waking dream, he gathered them in his arms, kissed and hugged them both. And smiled. Sweet dreams.

CHAPTER 12

Alaska

A tremor in the ground startled him awake. The tent shook vigorously. Earthquake? His eyes fluttered open just as something wet and slimy smeared across his forehead. McClure jerked his head backward and found himself staring at a big mottled gray nose with quivering nostrils. He scrambled backward in alarm and the creature flinched at his sudden movement, its eyes bulging wide. Huge eyes. What the hell? Dave sensed motion all around, shuddering quakes in the ground, the tent shimmying as something large brushed up against the sides. Again the creature shoved its head toward Dave, blinked its eyes and swiped its long tongue across his forehead.

McClure's lips opened in whisper. "Hey, what're you doing, huh? I ain't no lollipop, fella. Gees."

At the sound of his voice, the beast snorted, spraying mucous all over his face.

"Damn, thanks a lot!" He reached up and wiped his face on his sleeve in disgust.

The creature grunted, finally pulling its head out of the tent. Dave cautiously edged forward and poked his head out through the flap. The animal loomed over him, still looking at him curiously. It made a moaning grunt that sounded like a cross between a horse and a goat. Another creature behind it bumped it forward with a snort. Wide awake now, McClure wiggled closer to the flap and fully unzipped it, but didn't yet dare to crawl out. His tent was surrounded by hooves, lots of hooves split with four large toes, thumping on the tundra like a symphony percussion section run amok, kicking up clumps of snow as they went. The ground shook from the sheer numbers of them and nearness to the tent. Caribou!

Dave pushed himself to his knees and gazed out in astonishment. A herd of caribou was passing through his campsite, way too many of them to try to count. Most had antlers but quite a few didn't. He wondered briefly until he remembered Ed Weiss saying the females had antlers too and kept them until spring to protect their calves.

A few caribou turned to look at him as they moved along, bumping up against each other and milling around the trees, but they now mostly seemed disinterested. Apparently, they'd decided he was not a threat. But, yuk! He'd been slimed! He guessed at least a hundred of them were meandering past, weaving their way through the early morning under a sky just beginning to lighten. Part of... what did Ed Weiss say? Oh yeah, the Nelchina herd. What a sight.

As the last of them tromped from his campsite, he crawled out of the tent to watch their departure, a look of amazement on his face. They twisted their way leisurely through the trees heading south, making a gradual turn to the southwest. Mesmerized by the sight, he watched the last of them disappear. Finally, Dave grabbed his water bottle and splashed some water on his face, again wiping the stickiness off with his sleeve. Ugh. Nearly French kissed by a caribou!

When he got back to Valdez, he would not skip telling Ed Weiss about this closest of close encounters. Curious critters indeed. He had to admit though, these were gorgeous animals, just beautiful with their large eyes, plush coats and majestic antlers. His eyes skimmed the campsite. All the snow in sight had been thoroughly trampled. Any snowshoe tracks from his trek up here yesterday were now gone, obliterated by the caribou.

He crawled back into the tent and slipped on his boots and parka, then stood looking up at the sky. Despite the growing light, it didn't look as though the sun would be out today. Clouds, lots of them, and he could smell heavy moisture in the air. Snow. The question, of course, was how much, just some light flurries, a trace, or a blizzard? It was early November, so he doubted a blizzard was coming in, but it certainly smelled like snow was on the way. He'd have to be sure to take his bearings today and take them frequently. But, any snow at all would be to his advantage as it would absorb sound, muffling any noise of his trekking.

McClure pulled out the Sterno stove and lit it. He put some water in the metal cup and set it on the stove, then mixed in some instant coffee. Today would be a long day, so another pack of that Mountain House scrambled eggs and bacon sounded good. He rummaged in the backpack for the package. As he sipped his coffee, he reached under the tent for some clean snow and stuffed it into his water bottle. He checked his parka to ensure that the compass, some beef jerky and crackers were there, at least enough for lunch.

Breakfast was great. As he squatted outside the tent and sipped his coffee, he could feel the air thickening. Yeah, it was going to snow. By the time he'd cleaned up the utensils and packed everything away in the tent, it started. With hardly any wind to push them along, light flurries zigzagged their way down through the trees. McClure raised his face to the sky. A few flakes settled on his nose and forehead, tickling as they melted. This place was truly a wonder.

The birds were unexpectedly back too. A bunch of

chickadees and the gray sparrows flitted around, dancing in the tree branches. The whole scene was as picturesque as a Currier and Ives print. The lacework of quivering flurries wove through the birches and aspens in their journey downward, finally clumping together in anonymous crowds on the ground. Dave gazed in wonder at the astonishing beauty of a snowfall serenely anointing the surrounding woods.

As magnificent a scene as it was, he knew he needed to get underway. He reached into the tent, grabbed the rifle, zipped the tent flap closed and tied it off at the bottom. His snowshoes were strapped on securely and he telescoped the trekking sticks. As he slung the rifle over his right shoulder, barrel downward, he noticed how incredibly light he felt without the backpack. That meant today would be fairly easy compared to the last couple days of carrying full gear. Even better, his hike would be mostly downhill for at least the first half the day.

The snowfall continued to be just a scattering of flurries. He spun around 360 degrees to get his bearings in the campsite. To start, he'd travel more to the northeast for a short while, then fully to the east, and eventually south and westerly. He pulled his goggles down over his eyes. Even in the flurries, it was surprising how incredibly clear the polarized lenses made the surrounding forest.

McClure set off to the northeast. He was getting this snowshoe thing down pretty well. The rhythm of his breathing and a relaxed pace came easy now. The air was frosty. He guessed the temperature had dropped down to about ten degrees. But as he trekked through the wondrous landscape, it wasn't difficult to ignore the cold. Time ticked away seamlessly.

There was so much to observe that... Dave stopped abruptly, staring at the snow covered ground in front of him. Grasping both sticks together in his left hand, he knelt down and laid his right hand into an imprint in the snow. No mistaking it. A bear.

A grizzly for sure, and a huge one at that. Probably one of those two brother grizzlies that Ed Weiss mentioned, the ones that never seemed to hibernate through a whole winter. The ones Jack Stroud had seen quite often, almost every time he entered the wilderness to the northeast of Valdez.

The grizzly's tracks crossed his path, heading due south toward the ice field. McClure bent over and examined the edges of the tracks closely. They were still crumbly and hadn't yet crystallized, which meant they were relatively fresh, probably within the last eight hours. He swung his rifle up onto his lap and drew back the bolt. A round was in the breech. The safety was on. He stood, slung the Weatherby over his right shoulder again, and looked to the south.

Well, if anything could wake him up out of his recent woodland musings, this was it. A genuine, carnivorous grizzly, probably weighing in close to 1,500 pounds. A monster of a bear. And, not all that far away either. Dave looked at his watch... 11:30 AM. It was about time for him to also swing south to make the wide loop he had planned to take... and right behind the grizzly.

He shifted his course to the south, weaving through widely spaced trees. The forest was opening up, mostly black spruce again now. The bear tracks veered away closer to more dense woods to the west. That was a bit of a relief, at least for now. His own path was skirting the edge of that thicker forest, stands of tightly packed trees lying off to his right. The grizzly was in there somewhere.

The snow flurries were intermittent and had become very light. He estimated the temps had risen to 20 degrees. A regular heat wave. Well, no better time to take a lunch break. Dave moved to his right and entered the tree line, then took out his knife and cut some spruce boughs to sit on. He plopped down facing the woods, his back to the more open area behind him.

He didn't want the surprise of a grizzly charging him from the woods. Facing this way, he'd have at least some warning.

He wasn't all that hungry, but the jerky and crackers did feel good in his gut. He raised his water bottle and took a swig to wash it all down, then drank about half the bottle in just a few gulps. Without hesitation, he packed more snow in the bottle. Staying hydrated in the wilderness was critical even in winter.

As he stepped out from the trees, something caught his attention. He turned his face into the light breeze moving north across the tundra from Controller Bay. Smoke… wood smoke. The smell was faint, so wherever it was coming from, it couldn't be too close. But somewhere to the south or southwest, someone had a campfire. McClure had a pretty good idea who he was, and he would need to slow his pace and acutely expand his awareness to any movement in the woods. The light flurries had softened the surface of the snow, dampening the sound of his snowshoes. Stealth was now highly important.

He hadn't traveled the length of two football fields when he saw bear tracks again cross his path, this time traveling west again. The bear must have crossed back north of him, but he was certain it was the same grizzly. The bear was making a loop as well. He scanned his eyes along the woods. Nothing. Perhaps the smell of the smoke was spooking it, making it travel erratically, and it had now reentered the forest. McClure figured he'd better go even further south before he too swung right, heading west into the woods. The last thing he wanted to do was to run into a grizzly, especially one that big and probably looking for something to eat.

By 1:00 PM, he started curving west into the woodland, where black spruce grew very closer together. It was impossible to guess where the grizzly was, but he thought it likely the bear was at least six hours ahead of him. Dave hoped it had continued due west and would then be far away by now. But, not far enough to disregard its presence.

The smell of smoke was markedly stronger here. The ground was rising to the southwest and despite the dense growth of trees, McClure sensed he was approaching the edge of something, perhaps a ravine or even a valley of some size. The woods appeared to end about thirty yards ahead. The flurries had stopped, but the air was still saturated with moisture so he knew the snow could easily start up again. He glanced to the sky and saw trails of smoke drifting up over the tops of the spruce trees ahead. Whoever they were, they were here. And close.

Dave leaned the rifle against a spruce and sat down to remove his snowshoes. The snow wasn't that deep, and he needed stealth now. He set the sticks by the snowshoes and grabbed his Weatherby, then rose to his feet. Hunched over, he cautiously stepped forward until he was maybe ten yards from the edge. He eased down to his knees, and with his rifle laid across his forearms, began crawling forward through the snow. Gradually, he worked his way under some low hanging spruce branches, squirmed his way up to the edge, and peered over it.

It was a deep ravine, eighty feet or so down to the bottom. Probably an old stream bed. His eyes followed the wispy trails of smoke down to their source. He could make out at least four tents lined up to a fire pit, and then one larger, open tent. Groping down the front of his parka, he pulled out his binoculars and popped off the lens caps. Close-up and crystal clear.

He saw men milling about in puffy white parkas and matching pants. With their rounded bodies and stubby legs, McClure chuckled quietly to himself imagining them as large, animated snow men. Snow men should have pieces of coal for eyes, but these snow men wore dark colored goggles. They had apparently chopped down a lot of the birch and aspen that filled the ravine to clear the campsite and also for firewood, given the pile of wood he saw stacked at the edge of the clearing. There were two tents on each side about ten feet from the fire and with about five feet of space between them.

The binoculars zoomed in to the larger tent, its flaps tied back to the poles. He counted three snow bikes inside with what looked like spiked tires. All three were two-seaters. A couple of the men standing near the fire wore some kind of weapons slung over their shoulders, stubby rifles or carbines. And all of the men he could see were wearing pistol belts. Resting on his elbows, McClure steadied the binoculars and carefully counted the men as they walked about, chatted, entered and exited the tents. After nearly ten minutes of surveillance, he was fairly sure there was a total of eight of them in the camp. Were these the ones who had murdered Jack Stroud? There were certainly enough of them to match up with all the footprint indentations that he'd found in the snow around Stroud's body and had appeared to head away in a group to the southeast.

Eight men all dressed exactly alike in white winter parkas and matching pants. Perhaps he had inadvertently run across an international geologic expedition conducting tests on the tundra of the Alaskan wilderness. Geologists who wore assault rifles slung over their shoulders and semiautomatic pistols on their belts? No, this was something else entirely. And when the hell had McClure ever seen a bunch of hunters packing into the wilderness and all wearing the same get-up? Never. These men were dressed exactly the same right down to their boots, like... uniforms.

Dave looked south toward the bay and observed that the top of the ravine dropped as it ran toward the shore. Black spruce trees were packed along the ridge line the whole way. Okay, he might be able to get closer. If he could get close enough to hear them speak, he might be able to discern who these men were. McClure had a fair understanding of several languages...

French, German, Russian and Japanese.

What would be even better was to get a closer look at their weapons so he could take an educated guess at what kind of group they were. From what he'd seen so far, this had to be a military unit of some kind. He'd had weapons training in both the CIA field operations school and his ten years with the FBI. He'd been away from the 'business' for a year, but that wasn't long enough for him to have forgotten all that training.

He scooted backward away from the edge, careful not to disturb the overhanging branches. Once below the ridge line and out of sight of the camp, he shouldered his rifle again and crawled on all fours to the south, pausing every now and then to listen. As remote as this area was, it wasn't beyond the realm of possibility that the snow men might still have perimeter guards out. But as he crawled along, he neither saw nor heard any. There were no guards, perhaps because he was east of the camp. Nothing further out this way but ice fields and rocky wilderness. It took about twenty minutes to get where he thought he wanted to be, closer but not too close. As he inched back up to the edge of the gorge, he saw that he was now only about forty feet above the bottom.

He pulled out his binoculars and focused on two snow men, talking and smoking near the southern perimeter of the camp. Behind those two, another two men were kneeling by the large open tent and next to a box covered in what looked to be white canvas or nylon. The cover was partially zipped open and one of the men held a phone-like device at the end of a coiled cord. A portable transmitter? Something akin to a combat net radio. But to transmit what and to where? What could possibly be close enough to the camp?

McClure swiveled the binoculars back to the two men smoking and pushed out the zoom lens to zero in on their rifles. They looked like PP-2200 submachine guns. What the hell?

The PP-2200 was a full auto, Russian-made assault rifle that rapid fired 9mm parabellum rounds. It was intended for close quarter-combat and used mostly by Russian special operations forces – Spetsnaz. Now that was interesting.

Dave lowered the focus of the binoculars to the men's pistol belts. He could make out the grips; they were very distinctive. He'd seen these pistols before and had even fired them on the FBI range... Grach pistols, MP-443s, also 9mm and Russian made. This set up was beginning to put him on edge.

He crept further forward and stretched himself out to the cliff's edge, keeping well under the cover of the spruce while straining to hear them talking. The ravine actually helped by funneling their voices upward and the sound carried well in the cold air. McClure felt the hair on his neck stand on end as a wave of gooseflesh swept over him. His eyes bulged, and he shook his head slowly in disbelief. He could clearly understand what they were saying. It was Russian.

CHAPTER 13

Israel

The drive back to Tel Aviv was dead quiet, almost surreal. They didn't talk at all. In contrast to the bright sun, the interior of the car felt dim and subdued. Raphael must have looked over at her at least fifty times. Meira drifted in and out of sleep, punctuated by episodes of whimpering, almost crying. Her face was stained by marks of tears. A policeman had helped him half-carry her for about a block until she could walk on her own. She was in no condition to drive, so they'd left her car in Jerusalem.

Raphael told her he was taking her home and she seemed grateful, but that was the last thing she'd managed to say. The incident had shaken her badly. She had killed a man, a young man, someone her age. At some point, he felt he would have to coax her to talk about it, but for the last hour, he had just let her sleep.

"Meira?" He leaned slightly over to her.

"Yes?" Her eyes blinked.

"Meira, where am I taking you? Where do you live?"

Meira murmured an address off Highway 4. Her eyes closed again.

Meira lived in Bnei Brak, a little suburb just a short distance north of Tel Aviv. As the exit sign approached, Raphael swung the car around the loop off Highway 1 and headed north on Highway 4. Just minutes later, he pulled into the complex. The apartment house looked new, a bleached white exterior and what appeared to be tinted glass, thermal windows. Raphael pulled into a parking place near the entrance and gave Meira a gentle shake. Her eyes opened, startled.

"Meira, you're home. Your apartment house. You okay?"

She turned her head slowly toward him. "Home already? Rafi, I must have really been gone."

"You were... in and out. I kept an eye on you. That was quite a situation back there and you performed fantastically. You just need some rest."

What time is it?"

"It's about 12:20. I took it easy coming back."

"It's only lunchtime. I can't sleep now."

"Well, you can decide that when you get up to your apartment. Think you can walk okay now?"

"Yeah, I'm okay. Rafi, thanks for this. I don't think I would've been able to drive."

"That's right, you wouldn't have been. That was easy to see. Tomorrow I'll drive you back to pick up your car."

"Thanks again. Please come upstairs with me. I'm on the fourth floor. You know, just in case."

"Sure."

The building sported a nice lobby with some kind of polished stone flooring and casual seating scattered around. She held his arm as they stepped out of the elevator on the fourth floor and walked the few doors down to her apartment.

"Please, come in." Her hazel eyes beckoned.

"No, not now. I've got to get over to the center and send out some preliminary notifications about our involvement at the site of the bombing. And you need to rest, take a warm shower. But, can I pick you up tomorrow at 7:30? Is that too early?"

"Rafi, please..."

"Not now. Some other time, I promise."

"You're building walls."

"Meira, some other time, okay? Get some rest." He bent over and gently kissed her forehead.

"You're a poop."

"I know. See you tomorrow." Raphael turned away, waving his hand behind him. No way was he going in her apartment now. He could understand why she wanted some company, but it would be a bad move on his part. Meira was hurting and vulnerable. No, not now.

He drove to Mossad Center, and went in the first floor cafeteria to grab a soda and a chicken salad sandwich. He rode the elevator up to his office, plopped down at his desk to log onto his desktop computer and then, between bites of the sandwich, started a series of email notifications about the attempted bombing in Jerusalem to the Mossad Operations Center; to Taavi Perutz, the director of Mossad; to General Ariel Goren; to Yakov Sivitz, deputy minister of defense; and to Ethan Kirschner, deputy director for Collections.

Raphael highlighted Meira's quick actions in analyzing the developing situation and taking down the would-be bomber. In the ensuing violent, close-quarter combat, she shot the man twice in the head, preventing him from detonating the IED. In so doing, according to police on the scene immediately after the incident, Meira had saved hundreds of lives.

Meira was okay, roughed up a little with some bruises on her throat where the man had grabbed her, but no significant injuries. Raphael had taken her to her apartment to rest and recover from the incident, and he would pick her up and bring her in tomorrow morning at about 8:00 AM. Raphael closed the email with a recommendation for Meira to receive both the Mossad Plaque of Excellence for heroic bravery and the IDF

Medal of Courage for gallantry at the risk of her life. He washed down the remainder of the sandwich with his Coke, rereading the message before he clicked "send."

He walked over to the coffee machine, poured himself a short cup and stood by the window looking out to the Mediterranean. Nice day out. A few white clouds were drifting in from the sea, but overall it was a vividly blue sky. *Such a contrast,* he thought, *the sheer beauty of Israel and the violence taking place almost every week.* If it wasn't rockets flying in from Gaza, it was some crazed Palestinian bomber like the one today. He shook his head and sipped his coffee, wondering if it would ever end.

"Raphael," Leah, the department secretary, interrupted his reverie. Leah was 60, increasingly thinking about retirement and almost daily asking anyone within earshot their opinion of that possibility.

He turned from the window. "Yes, Leah?"

"General Goren is on the phone for you."

Raphael hurried to his desk, set down the coffee and picked up the line. "Yes sir? This is Raphael."

"How is Meira?" The concern in the general's voice was palpable.

"She'll be okay, sir. It's mostly mental at this point. It shook her up, killing that guy. But you wouldn't have guessed that would happen by the way she raced out from our table and tore into him. He was behind me… I never saw him. If it weren't for Meira, I wouldn't be talking to you now."

"Interesting about her. I've seen it before, though. Sometimes you really don't know what someone is capable of until the moment comes when they have to act and act fast. Meira's a warrior. She'll get through this, come to grips with it and settle down. I'd like you to bring her into my conference room tomorrow morning at 10:00 AM?"

"Certainly, sir. We should be in the building by 8:00. I want to talk to some of her Collections people in the technical area."

"About the Krasnoff and Madani matter?"

"Yes, sir."

"Very good, Raphael. Stay on it. That's the priority. See you tomorrow. Ah, if you can find a necktie, wear one?"

"Yes, sir. Shalom." Raphael set the receiver down and sighed. A necktie. This must be big. Unable to concentrate on work any further this afternoon, he went home, had a beer and went to sleep. His dreams were surreal repeats of Meira sprinting from the table, taking the man down so violently and efficiently, and finally shooting him.

CHAPTER 14

Israel

When Raphael's silver Passat rolled through the parking lot the next morning, Meira was standing outside the apartment house. She tugged open the door and slid in. Before he could say 'good morning', she turned to him.

"Rafi, thanks for yesterday. For everything. And, for being so much wiser than me."

"Meira, you did something very heroic... you displayed so much courage. You were badly shaken, that's all. You feel better this morning, right?"

"That's not what I mean."

"I know. Let's just leave all that untouched for now. By the way, after we talk to your people in Collections, General Goren wants me to bring you by his office at 10:00 AM?"

"Sure."

"Did you sleep well?"

"Dreams off and on all night long. I kept seeing that man. In the dreams, I never reached him in time. The bomb went off. I was blinded by the blast and felt a searing heat on my face, my arms and legs. Then, I'd wake up. Over and over. But I got some sleep anyway. I'm okay."

"Well, truth is, I dreamed about you taking him down and shooting him, too. You got him, Meira. The bomb never detonated. Your dreams were pure fear overwhelming your subconscious, fear of not getting to him in time. It… it sounds very normal, really. Other than that, you're fine, right?"

"Yes. Again, thanks, Rafi. By the way, what's with the necktie and sport coat?"

"General Goren called me yesterday afternoon after I dropped you off, asking how you were. He said I should dress up a little for our meeting."

The ride to the center was a brief 15 minutes. Both guards in the lobby smiled and nodded respectfully to Meira. Word travels fast. After grabbing coffee and a couple bagels, they took the elevator up to the Collections department on the second floor, and wound their way through a labyrinth of cubicles to small conference table.

"Micha, can you come out here, please?" Meira called out, her eyes scanning the open bay room.

"Meira, is that you?" Micha appeared out of the maze, smiling broadly. "How are you? We all heard about yesterday. We're so proud of you!"

"Thanks very much, Micha. I'm still trying to decide in my own mind if what happened was actually real."

"Sure, I think that's how most anyone would feel the day after something like that. Ah, how can I help you?"

"Please, have a seat. This is Rafi Mahler. He's in Metsada, special operations. We'd like to ask you about nuclear weapons, warheads." They all sank into seats around the table.

"Okay, hi. Nuclear weapons?"

"Yes," Raphael replied. "The American ones, primarily. The ones they put on their missiles, ICBMs."

"My, this is an interesting subject. So, warheads on ICBMs, American strategic ballistic missiles. May I assume then that you're discounting the SLBMs and SLCMs, ah, that's Submarine Launched Ballistic Missiles and Cruise Missiles, and the ALCMs,

Air Launched Cruise Missiles, and of course, the bombs?"

"Yes, just the ICBMs." Raphael replied.

"Well then, that really leaves only one delivery vehicle. That's the American Minuteman III missile. The technical weapon system name for it is the LGM-30. It's a three-stage, solid fuel missile."

"Good. Where are the launchers for the Minuteman III missiles?"

"They're all in underground silos, Raphael. Large missile fields of underground silos. Only three locations now... F.E. Warren Air Force Base in Wyoming; Malmstrom AFB in Montana; and Minot AFB in North Dakota. What the Americans call their northern-tier bases, all to the far north in the U.S., and all very well protected by Air Force security forces personnel. I've heard that since the U.S. Air Force took some heavy personnel cuts, they may have even installed microwave fencing around the missile fields. It's very painful to approach one of those microwave fields, and if someone gets too close, it kills them, and very efficiently."

"All in underground silos?"

"Yes. And underground command centers with, I think, two officers, controlling about ten missile silos."

"I see. It all sounds very secure, impenetrable."

"I would say so, yes. I've never heard of a single breach taking place at any of their missile fields."

Raphael rubbed his chin and glanced at Meira. She looked puzzled, wondering where Rafi was taking this line of inquiry. He looked back to Micha.

"Micha, what can you tell me about the warheads in these Minuteman III missiles? A single nuke?"

"Some are single. But most are MIRVs, that's Multiple Independently-targetable Reentry Vehicles. Three nuclear reentry vehicles on the warhead on each missile."

Raphael's face paled. "You said MIRVs."

"Yes."

"What is the yield of these warheads, ah, you know, kilotons?"

Micha's lips curved upward in a muted smile. Opportunities to demonstrate the breadth of his technical knowledge were rare. "About 450 kilotons each. The way it works is that after launch, the missile travels through exospheric space to a general target area. It maneuvers into a ballistic trajectory that sends a reentry vehicle with a warhead to a target, releasing a warhead on that trajectory. The missile then maneuvers to a different trajectory and releases another warhead, and so on."

"MIRVs. 450 kilotons each." Raphael felt queasy. Meira looked him over with concern.

"Yes."

"Sounds like a devastating weapon system. Pretty fool proof."

"It is, really. When you consider that ten or so missiles with three MIRVs each may be approaching a certain geographical area... well, once the nose cones open and deploy those reentry vehicles with their warheads, it plays hell with any land-based antiballistic missile system."

"Dammit!"

"What'd I say?" Micha's eyebrows rose.

"Nothing, I'm sorry. Micha, thanks very much. But, I just don't see how...? Please, where's your men's room?" Raphael stood.

"Against the wall over there." Micha pointed.

Meira stood as well. "Rafi?"

"I'll be right back." He walked away, shaking his head and muttering. Meira and Micha stood staring at each other, looks of apprehension awash on their faces.

Raphael shoved open the door, walked to the sink and bent

over. Staring into his reflection in the mirror, he splashed cold water on his face. He needed to regain his composure. From Micha's answers to his questions, things were looking worse by the minute. When he exited and closed the door of the restroom behind him, Meira was standing and waiting. Micha had gone back to his desk.

"You okay?" Meira asked.

"Yeah. But we need to talk to General Goren as soon as we can, at the 10:00 meeting if possible. Why don't we head up there now?"

"What is it?" Meira persisted as they approached the elevators.

"Well, what Micha just told us tracks perfectly with what Dani believes he saw Krasnoff say to Madani. But the name wasn't 'Marv' or 'Merv', not another man, it was MIRV. And the number '450' that Dani mentioned correlates with the yield of the reentry vehicle warheads. It matches up too well to ignore it."

"What are you going to recommend?"

"If this whole thing… you know, the meeting in Paris, the deployment of Krasnoff's GRU brother and a Spetsnaz team to Petropavlovsk in Kamchatka, a possible rendezvous with a submarine, if Dani read Krasnoff's lips correctly, if all this is about a nuclear weapon, a warhead, then logic tells me it's a plan to steal an American Minuteman III MIRV warhead. That's a lot of 'ifs', I know, but it's the only thing that makes sense to me right now. But…" Raphael paused in thought.

"But what?"

"But, if these missile fields with underground silos and command centers are so secure, then how the hell can they get to one to steal it? That part makes no sense at all."

In minutes, the two walked into General Goren's outer office. The secretary smiled, nodded knowingly, and gestured toward the conference room. As soon as they stepped through the door, Meira stopped dead, staring around the room. The room overflowed with people from all over the government, even Taavi Perutz, the Mossad director himself. They all smiled at Meira as General Goren motioned to her to join him at the front of the room. Meira shot Rafi a worried look.

He gave her a gentle nudge. "Go on, Meira. This is your time." He squeezed her hand and left her side to stand with the others.

Meira gasped and raised her hands to her mouth in shock as General Goren enveloped her in a bear hug. She looked at Rafi, who was grinning broadly, and she almost choked with emotion. Blinking back tears that sprang to her eyes, she noticed heads turning and people standing at attention for someone behind her. But before she could turn around, Ari spoke, his voice formal and official sounding.

"Ladies and gentlemen, let us welcome Ms. Meira Dantzig, an officer of Mossad and a very courageous young woman." Cheers and clapping filled the room.

General Goren continued, "And, please welcome our distinguished guest, General Alon Dresner, Commander of the Israel Defense Forces!" The clamor grew to a raucous din. "Please, General Dresner." With a graceful gesture, General Goren stepped aside.

General Dresner was a tall, lean man of at least 6′3″. His bronzed face and build alluded to his athleticism. A well-manicured moustache was the only distraction in his appearance. He smiled at Meira and raised his hands for quiet.

"Ladies and gentlemen, thank you for the warm welcome. I just happened to be in the neighborhood when Ari called and asked me to come over." He looked again at Meira, whose eyes were brimming. "Truth is, I wouldn't have missed this for the world. Come here, child."

Meira stepped cautiously toward him and he also engulfed her with a hug, keeping a hand lightly on her shoulder. The room again erupted in cheers.

"I'm here today to present Ms. Meira Dantzig, this daughter of Israel, with the Medal of Courage for her gallantry yesterday in Jerusalem. At the risk of her own life, she saved the lives of at least a hundred citizens. It is a rare occasion that we present this IDF medal to anyone outside the military, but without question, Meira's selfless actions merit it."

A young lieutenant stepped forward with the medal case. General Dresner gingerly pinned the medal to her blouse. He stepped aside, clapping, and once more the room rang with applause. A tear slipped from Meira's right eye and slid down her cheek. General Dresner smiled and pulled a handkerchief from his jacket pocket, reached over and gently captured the tear.

"Taavi?" General Dresner nodded.

Taavi Perutz, Director of Mossad, came forward. About 60, balding and developing a slight paunch, Perutz wielded tremendous power both with the prime minister and the members of the Knesset. He was also without a doubt the best-dressed man in the room, exuding his usual panache. He stepped up to Meira and kissed her cheek.

"Ladies and gentlemen, I am pleased to present our own Meira Dantzig with the highest award Mossad offers, the Plaque of Excellence, for her heroic bravery yesterday in Jerusalem. Meira, we thank you so very much, and want you to know how happy we are that you are such an important member of our Mossad family! Congratulations!"

The room exploded in applause and cheering. No stopping it now, Meira's chest heaved and the tears flowed freely. The applause pounded the walls and windows and echoed down the hallways around the entire floor. General Dresner motioned to General Goren and Ari came forward and hugged her again.

"This is no time for tears, Meira, but rather a time for joy and even prayer. Ladies and gentlemen, please congratulate Meira Dantzig!" The cheering and applause continued.

Finally, the general motioned for silence and the applause abated. "I would like to add at this point that it is no small thing to take another person's life. Indeed. A photo flashed on the screen at the front wall. But this young man, Naseer Yassin, had already taken the lives of many and intended to do so again yesterday in Jerusalem. Yassin was fully committed to the terror objectives of Hamas."

The general continued, "We know now from fingerprints that it was Yassin who killed 24 people last summer in the bombing outside the town of Arnona. This young man, an educated man, was also very much a vicious killer. This we must not forget. For all of us, and especially you, Meira, I would like to quote a few verses from the Tanakh, Ecclesiastes 3:1-8."

Ari looked around the room before beginning. "*There is a time for everything, and a season for every activity under heaven, a time to be born and a time to die, a time to kill and a time to heal, a time to weep and a time to laugh,'*... and also... 'a *time to love and a time to hate, a time for war and a time for peace.'* Praise be to God for His mercy on us, on our people, our nation, and for His gift of children such as we honor today, Meira Dantzig."

Applause rang up again, however more restrained this time as more than a few eyes were wet around the conference room. Meira was hugged and kissed more times than she could ever remember as the attendees passed, congratulated her, and left the room. She was overwhelmed with emotion, her eyes red and cheeks wet with tears. General Dresner, Taavi Perutz, Yakov Sivitz, and Ethan Kirschner each took one last opportunity to shake her hand before they left.

Finally, the last well-wishers had left, leaving Meira and Ari standing at the front of the conference room. He patted her shoulder as Raphael approached. "Meira, we are all so very proud of you, and you should be proud too. Words alone cannot fully express our feelings about what you did yesterday. And you, Raphael, you should be proud too."

"I am."

"Good. Raphael, you said you both wanted to talk to me. About the operation?"

"Yes sir. Meira and I both feel it may be urgent. You and Mr. Perutz may even want to discuss this with your contacts at the CIA, today perhaps. At least, that's our recommendation."

"All right, let's go into my office. Incidentally, you should know that Viktor Nechayev met with Leonide Krasnoff in Amsterdam."

"Viktor Nechayev came out of Russia to visit with Krasnoff himself? This has to be significant, General. That's very disturbing news."

"I agree. It is significant."

"Did surveillance get anything useful?"

"No, nothing. Our team couldn't get close enough. They met at a place called the Harbor Club on a little extension of land jutting out into the harbor. Much too isolated, and the meeting took place before the lunch hour. The site was well selected, and security was heavy. The team would have stuck out, so they aborted the op."

"Too bad."

"Yes. Come to my office. You two can tell me what you've learned."

CHAPTER 15

Alaska

The moon hid behind a thick cover of clouds. The snow had stopped but the sky was still low and sullen. Even so, the snow provided a remarkable amount of reflective light, making it easy to follow his own tracks back to the campsite. As he trekked, his mind buzzed with questions. What were Russians doing in south central Alaska? From the looks of them, probably Spetsnaz special forces. How the hell did they get here? And why? And what the hell was he supposed to do about it? What *could* he do?

One thing was for sure, he'd get up tomorrow at o'dark-early, trek back to their camp and watch them for a while. See if he could figure out why they were on U.S. soil and what they were up to. The trip would be a lot faster since he now knew the location of the camp.

McClure stopped suddenly and stooped over, glaring at the snow no more than a few feet in front of him. Bear sign again, the tracks crumbly and clear in the snow. These tracks were fresh. He continued tramping forward, but more slowly, his eyes darting back and forth alertly. It seemed a tediously slow pace. Eventually, he made the turn west toward the thick woods and his camp.

His tracks from early that morning passed between two black

spruce toward a large stand of birches just beyond. He paused, frozen in place, hearing the cold snap of branches. More snaps. Very distinct, and coming closer. A rush of gooseflesh seized him. Something large was moving toward him.

The horrendous jaw-popping roar shattered the silence and echoed in the surrounding woods. Stomping into the open from behind the spruce on the right, looking like some prehistoric behemoth, the grizzly was on its hind legs, swinging its head back and forth. Another pulsing roar followed the first, dwindling off into a snarling growl through the bear's jagged fangs. McClure's mouth dropped open in shock. He was paralyzed by the sudden assault barely twenty feet in front of him.

The bear was gigantic, towering he guessed nine feet high. Dave tried to free his mind and think. If he could drop his sticks to the snow and swing the rifle off his shoulder… he might get off one shot before the grizzly was on him. But no, he wouldn't have a chance. The gamble of hitting a neural area that would stop the bear with one shot was slim to none. No, the grizzly would kill him. On an impulse, he raised his metal trekking sticks high over his head and began clacking them together rhythmically… a steady metallic clacking of the sticks. *Click, click, click!*

"Well now, aren't you a fine looking fellow, you sonofabitch." McClure spoke in a steady monotone voice, rapping the sticks repeatedly above his head. *Click, click, click!* He stood his ground, not backing up, not giving up, trying desperately to present a display of dominance.

"I'm telling you, I am so pleased to run into you tonight. Yessiree, I am just so happy all to hell to meet you, Mr. Grizzly, or perhaps you would prefer, ah, Bruno or Brutus? Really, there's nothing else I would rather be doing. You know, I've been admiring your paw prints in the snow all day." McClure struggled to keep his voice steady and uniform. *Click, click, click!*

The bear's growling *huffs* dropped off to a low grumble. Its huge paws clawed the air as it swung to and fro on its hind legs. The distracting metal sticks swaying slowly back and forth and clicking, almost seemed to mesmerize it. The grizzly suddenly dropped down on all fours and McClure's eyes bulged. Was it going to charge him? It was a monster even on all fours, at least five feet high at the shoulders.

Click, click, click! "You know, you wouldn't like me at all. Really. I'm just a lean flank steak, stringy, hardly anything on my bones. But Mr. Bear, I can show you where you can find a whole bunch of fat, juicy-ass Russians just south of here. You'd like them. Oh, yeah. Whaddya say, huh?" *Click, click, click!* His arms felt heavy, straining to keep the rhythm of the sticks going above his head. McClure thought he was going to puke any minute.

The grizzly swung its head back and forth, its eyes boring into McClure's. It took a violent sweep at the snow at its feet and blurted a snarling *huff*. Its immense head looked as though it weighed 200 pounds by itself. Low grumbling grunts continued, the bear almost moaning now.

Then one abrupt high pitched roar. McClure kept banging the metal sticks. *Click, click, click!* Grunting, the grizzly turned away and plodded just beyond the two black spruce, then swung its head back toward McClure and *woofed* one last time before heading north at an easy lope. Skirting the forest and turning east toward the open meadow, it was gone in seconds.

"Damn."

McClure's arms fell limply to his side, letting the sticks drop to the snow. He collapsed to his knees, the rifle sliding off his shoulder. His face paled and he could feel his stomach churning. He lurched forward on his hands and knees, and vomited. There wasn't much left in his stomach after the light lunch, but he couldn't stop the dry heaves. He reached out blindly for some snow and rubbed it across his mouth.

After a few minutes, he gathered the strength to push himself up to his knees, found his bottle and rinsed his mouth out with cold water. Dave plopped backward, sitting in the snow and taking a few sips of water, swallowing slowly. He swiveled around to face the direction the bear had gone, and sat there for a good five minutes collecting himself.

"Damn. Now THAT was fun. Better than Disney World's *Wild* fucking *Kingdom*." He muttered. Actually, he, Anne and Michael had tremendously enjoyed their trip to Disney's Wild Animal Kingdom Lodge four years ago. It had been a blast. "Sorry, guys, I really didn't mean that," he apologized.

McClure picked up his rifle, brushed off the snow, and ran the bolt back and forth to be sure it was clear. Collecting the sticks, he forced himself up, then started off back in the direction of the camp. He glanced alternately off his right and back to his rear every few minutes just to be sure the grizzly wasn't following. Dave grimaced, noticing the uncomfortable feeling of having something wet in his shorts.

At the campsite 45 minutes later, the first thing he did was open the tent to retrieve the nylon rope from his pack. He felt he was far enough north to set a perimeter security alarm again. No one but him would hear the slapping of the stake up here. Next he also scooped out a fire pit to get a good fire going. With their own much larger fire pit, the Russians would never distinguish his smoke from their own.

After the fire was going and suddenly feeling the urge, he grabbed some tissue and his pack shovel, scooted under the nylon rope barrier and walked off ten yards or so beyond the camp to relieve himself. His .44 magnum pistol was still strapped on his belt. To his dismay, he found his shorts were soaked. On reflection however, he was pleased that the grizzly had only scared the piss out of him. Chuckling, he buried his personal *number two* deposit in the underlying tundra, heaping snow on top.

On the way back into the camp, McClure decided that he'd celebrate his survival of the night's horrifying encounter by breaking open the one package of beef stew he'd brought along. That, and coffee. And oh... he just remembered. Crawling into the tent, he rummaged about halfway down inside his pack, pulled out the bottle and held it up almost reverently. A half pint of Jack Daniels. Yes. He screwed off the cap and took a swig. Ahhh... so good.

He returned to the fire, suddenly realizing he was starved, but also unsure he could keep the stew down. The whiskey should help take any remaining shivers out of him. Truth was, that bear really did frighten the crap out of him. He remembered reading a book on surviving bear encounters and had always wondered what it would really be like to actually come face to face with a grizzly. Well, now he knew. And once was enough for a lifetime, thank you.

The stew was great. The coffee was delicious. Of course, it was the same coffee he'd had this morning, but something was different now. It was intensely enjoyable. Surviving the confrontation with the grizzly felt like he'd been granted a new life. Everything was different now, even the forest and the night air. By this hour, the temperatures had to have sunk down to near zero, but he didn't feel it. He only felt revitalized, restored and he mused about his new attitude while he cleaned his utensils and then buried the area with fresh snow.

Dave stoked the fire with new wood and stood up, taking a deep breath of cold Alaskan wilderness air before crawling into the tent. He slipped off his parka and boots, and zipped the front flap halfway down, positioning the revolver next to his shoulder. He thought that if he dreamed at all tonight, it would be of bears. Big bears. Monster bears. As he squeezed into the sleeping bag, he felt the sudden weight of the day press down upon him. Fatigue overcame him, and in just minutes, his rustling movements grew sluggish. McClure closed his eyes and slipped into a soft, soothing and dreamless sleep.

CHAPTER 16

Alaska

The loud, barking hoot of a snowy owl shattered the early morning stillness, followed by a shrill screech and flapping of wings. McClure struggled out of his deep sleep. His eyes fluttered open, and he pulled his hand out of the sleeping bag to look at the watch… 05:30 AM. He pushed aside the tent flap and peeked out. Pitch black and frigid. Still, he knew he needed to prod himself up, eat something and get on his way. He wanted to be watching when those Russian snow men crawled out of their tents.

After donning the boots and parka, he re-holstered the .44 and crept from the tent. The sky had cleared during the night, stars glittered overhead and the temperature had fallen off precipitously. The air was polar cold, he guessed below zero. Leaning over, he saw that again the last of the fire's embers had survived the night. He grabbed some dry, scruffy brush and pushed it down against the still glowing embers. Dry sticks and branches followed and the fire flared cheerfully.

The nylon rope around the circular perimeter was secure, the metal stake still in its notch. McClure pulled the stake off the tree and stepped around the perimeter to coil up the rope. He stuffed jerky and crackers into his parka pocket for lunch and made sure the water bottle was full, then tugged the rifle out of the tent and out of sheer force of habit, worked the bolt back and forth, ensuring a round was still in the chamber.

The snow in the pot was melting and almost ready for chicken soup and coffee. After pouring in the mix and stirring it to a bubbling finish, he sat in the darkness of the campsite lit only by the fire and slowly spooned his breakfast. The soup tasted as good as it smelled. The coffee was good too; he crunched a cracker between sips, his mind dwelling on his trek back to the south. He was confident that at this hour, the grizzly he'd encountered last night was most likely curled up asleep somewhere. Nevertheless, he sensed that today's trip would be chancy, dangerous even. He felt certain the snow men were the ones who had killed Jack Stroud. And they wouldn't hesitate to do the same to him. Despite the risk, he had to find out what they were up to.

McClure wiped the utensils clean, put them back in the pack in the tent and zipped down the flap. After strapping on the snowshoes, he scooped up an armful of snow, dumping it on the fire and watching it sizzle down to just steam and smoke. With the rifle slung over his right shoulder, he stepped out of his campsite into an arctic night. This time, he headed southeast right away, cutting a corner off his route to save time. He would eventually pick up his path heading south.

The moon had already sunk down below the trees, but millions of stars shimmered horizon to horizon, all sparkling against a velvet-black sky. His eyes were feeling the sting of the freezing air, so he pulled down the goggles, then zipped up the collar of his parka and tightened the hood. It seemed a wonder that the animals of this wilderness could survive year after year in such a harsh winter environment.

The trip south was relatively easy and uneventful. He saw no fresh bear sign, that in itself quite a relief. At 7:30 AM, he was approaching the southern, shallower edge of the ravine. The sky was just beginning to gray to the east. Once in the tree line, he slipped off his snowshoes and laid them on the ground with the sticks. Crawling to the edge of the cliff on his hands and knees, he rested himself under the spruce boughs, pulled out his binoculars, and scanned the campsite from north to south.

One lone man in a white parka and pants was crouched near the fire, stirring a pot hanging from a spit. Another larger pot was simmering next to it, probably some kind of breakfast stew. The aroma of fresh coffee wafted up the sides of the ravine, and his mouth watered. Within minutes, men were stepping out of their tents, stretching and moving toward the fire.

"Major Krasnoff, Captain Ivanov… come have some coffee!" The crouching man called. Another man began filling cups with coffee and plates with stew, handing them to his waiting comrades. Grunting with pleasure at the hot meal, the men settled around the fire to eat their breakfast. Then suddenly in unison, the men began to rise from their places as two more men approached from the northernmost tent.

The two waved their men back down. "Sit and enjoy your breakfast, men! You're going to need it. We have much to do today. Major Krasnoff and I will be briefing you shortly." The two accepted coffee and stew and joined the group.

So, those were the two officers, the team leaders. The one who spoke had to be the captain, Ivanov. The other, Major Krasnoff. All of the men were large, not one of them under six feet. And by their willowy, muscled movement, Dave could tell that all of them were in top condition. As they finished their meal, they went to a barrel to scrub off their plates and wash out their cups. They then lolled about the fire pit, waiting.

Captain Ivanov walked in the midst of the men, patting backs as he went. "Men, we are going to do this by the numbers. I would like each of you to return to your tents, strap on your arms, spare magazines and water. Then please line up here again in five minutes. I see puzzled looks on your faces. For now, I can only tell you this mission is of utmost importance to Russia. And, each of you, all of you officers, have been chosen to form this team, the best Spetsnaz team I've ever seen, to accomplish it. Major Krasnoff has assured me that we will all be briefed more fully later. We have two destinations today, and we will travel to them in sequence. I need three men for the first leg, who will take two snow bikes and the transmitter."

Senior Lieutenant Sorokin's brow furrowed. "Major Krasnoff, can you tell us where we are? None of us recognize this place. When we came ashore… I mean, it looks like the taiga, maybe near Kamchatka, but then that hunter a few days ago spoke English. Where are we?" The taiga was the large forested tundra in the most northern reaches of Russia stretching from Karelia in the west to the Kamchatka peninsula in the east.

"We are in Alaska."

"Alaska? We are inside the United States?"

"Yes. South central Alaska, to the southeast."

"What's going on, sir? Are we preparing for war?"

"No, not at all. This mission will actually work for peace. Men, what we will undertake and achieve on this mission will put the American president under great pressure, and work to the great benefit of Russia. That's why it's so important. That's really all I can say for now. As Captain Ivanov indicated, please return to your tents, and be back here lined up in five minutes. Go now!" The men gave quick shouts and ran for their tents.

Krasnoff moved toward Ivanov. "Great team, Alexi. You are to be commended. I am looking forward to working with you and your team."

"Major, you are a GRU intelligence officer. We are Spetsnaz. I and the rest of my men recognize and accept your leadership on this mission, without question, sir. But I urge you to appreciate the uniqueness of this unit. We do things differently, mostly to further the cohesion of the team. Since the creation of the first Spetsnaz units by Marshal Zhukov in 1950, we and others like us have now served our country for nearly 70 years. Each man on my team would give his own life to save that of his teammates. You should know this."

Vassily smirked. "And you should know, Alexi, all that really matters is your first statement. It is imperative that my leadership and direction to you and your men on this mission must not be questioned. If you understand that, we will achieve greatness in accomplishing it. Is that clear, Captain?"

"Yes, sir, of course." Ivanov was now confident in his initial assessment that the esteemed GRU Major Vassily Krasnoff was truly a self-indulgent asshole.

The men hurried out from their tents in full combat dress and lined up near the fire pit, PP2200 submachine guns across their chest. Captain Ivanov stepped forward.

"Senior Lieutenant Kolenka Mikulich?"

"Yes, sir."

"Junior Lt. Ivan Sokolov?"

"Yes, sir."

"Ivan, how's the shoulder, getting better? You're not wearing the sling?"

"It's coming along well, sir, the bullet went right through. I was lucky. I'm getting full motion back."

"Indeed you were lucky… that stupid American rushed the shot. "

"Senior Lt. Dimitri Sorokin?"

"Yes, sir."

"Junior Lt. Vladislav Bardinoff?"

"Yes, sir."

"Junior Lt. Stephan Lipovsky?"

"Yes, sir."

"Junior Lt. Nicolai Zarestsky?"

"Yes, sir."

"Good then. Team, do we accept our charge?" Ivanov took out his pistol and raised it in the air. "*Uraaa!*" He yelled, cheering to his men.

"*Uraaa! Uraaa!*" The men raised their rifles, grinning and cheering back. Even Krasnoff joined in and casually waved his own pistol.

"Senior Lt. Sorokin, you and Junior Lts. Bardinoff and Zaretsky will take two snow bikes. Strap the transmitter to one of the spare seats. Travel south to the shore, you should arrive there by 09:30 AM. The sub will be waiting for you. Transmit that we have canvassed the area and are now ready to execute the first phase. Request confirmation. Understood?"

"Yes, sir." The three men walked briskly into the large tent, where two began fueling the bikes and one retrieved the transmitter and strapped it securely on the back seat. Their weapons slung over their shoulders, they wheeled the bikes out. Sorokin and Bardinoff climbed onto the front seats and Zaretsky sat behind Sorokin. They revved their engines.

"Go men. Be back here by 11:00 AM. *Uraaa!*"

"*Uraaa! Uraaa!*" Everyone cheered again as the two snow bikes wove down the ravine toward the shore. Krasnoff, Ivanov and the others walked back into the camp and refilled their coffee cups, then sat on boxes around the fire. Ivanov looked over at Mikulich.

"Kolenka?"

"Yes, sir."

"You, Sokolov and Lipovsky, when you finish your coffees, fuel up the third bike and strap on the rail ratchet wrenches and tie lifters."

"Yes, sir." Mikulich answered. "Ivan, Stephan, come with me. We'll finish our coffees in the tent." They stood, nodded to Ivanov and Krasnoff, and walked off to the large tent.

Ivanov turned to Krasnoff. "This way, we'll be ready to go north immediately when the men return from the shore."

"Excellent. Alexi, I'm going to go review the maps again. Besides, I have something much better to drink in the tent. You're welcome to join me if you like. When we drink together in our tent, Captain, we are just Alexi and Vassily. Fine with you?"

"Of course, sir. Thank you." Ivanov replied, smiling. *What a great fucking honor,* he thought. The two men stood, dumped the remains of their coffee near the fire, leaving the cups. They walked back to their tent and disappeared inside.

McClure squirmed back from the ravine's edge, careful again not to rustle the spruce boughs. "Damn. What the hell are these bastards up to?" he muttered. "Okay, McClure, you better do something useful… I guess I'm heading south."

The sun had risen over the horizon into a clear blue sky. Temperatures would rise. Dave strapped on his snowshoes, picked up the sticks and stepped out of the thick stand of spruce, turning south. The sound of the snow bikes reverberated up and down the ravine. He could hear them clearly as he trudged along. The snow he was tramping on now was pristine, not a track in sight. The trees became more sparse as he went along. Rocks, ice and snow were increasing. He came to a slight rise and plodded upward another fifty feet before stopping. He hunched over, then sank to his knees and crawled to the crest.

Below the bluff, the bay was filled with floating chunks of ice and even some icebergs. He checked his watch… 09:28 AM. The shoreline curved outward toward the northwest. The snow men were already there, but he could barely make out the three of them and the snow bikes. He unstrapped his snowshoes and flattened himself in the snow, pulling out his binoculars. Looking through the lens, the fat snow men in their white parkas and the bikes were crystal clear. Two men had unstrapped the transmitter, uncovered it and set it down on the icy bank. The other, probably Sorokin, had his binoculars out and was scanning the bay.

Something to the right, just west of where Sorokin was standing, made McClure shift his binoculars in that direction. He saw two large, oval shapes that stood out in their pale ice and snow surroundings, stark white and maybe fifteen feet long each. Okay, what were those covers covering? Rafts, of course, two rafts.

He scrunched down in the snow, making mental notes. The transmitter and two rafts were key to the Russians, and accordingly, key to him as well. Without the transmitter, the Russians couldn't communicate with the sub. Without the rafts, they couldn't get back to the sub even if it was on the surface. They'd never make it in the frigid water and the dangerous ice floes. Unless the sub carried more rafts, which McClure doubted in those typically cramped quarters, the Russians could be stranded. No way to communicate and no way to escape.

He returned his focus to Sorokin and followed his line of sight, scanning out to the horizon but seeing nothing for several minutes. Then far out in the bay, the floes parted in a froth of water. An oblong shape like a humpback whale breached the surface. But absent was the tell-tale white spray.

The black form continued to rise but no fluke slapped on the water. This was no whale. McClure's eyes widened in recognition: the conning tower of a submarine. The rest of the black hull rose gradually, finally settling amid ice flows that slid along and bumped against it. As he watched, its antenna extended up from the tower.

Dave zoomed the lens to scan the profile of the conning tower. Most noticeable was what looked like a pair of large windows on the front. He'd seen this feature before. It was a trademark of Russian attack subs, and he'd never seen it on any other nation's subs. It really didn't seem to make a lot of sense to have windows situated like that on a conning tower, so more likely they were some kind of underwater acoustic sensors. Still, he recognized the tower from photos he'd studied in intelligence school.

This was a Russian sub, no doubt about it, a Schuka-B class attack boat and probably the latest model, the Akula III. Dave remembered that *'akula'* meant 'shark' in Russian. It wasn't overly large and probably was chosen for its maneuverability… to be able to enter the bay and come close to shore. This sub was probably how the Spetsnaz team got here. There was the outline of a large hatch on the side and just to the rear of the conning tower. That had to be how the rafts were launched and the snow bikes and other equipment were unloaded.

He shook his head in dismay. A Russian attack submarine in a bay east of Valdez, Alaska, and an eight-man Russian special operations unit on shore. Men who more than likely had murdered an innocent American hunter. Yeah, this game was getting more interesting by the minute.

None of the crew appeared on the conning tower, so the sub was most likely receiving transmissions from the men on shore. Naturally, they would need to be able to dive quickly if any aircraft showed up. He was surprised really that the sub surfaced at all, instead of just rising to antenna depth. But then, they had to have their reasons for this unusual operation.

The men at the transmitter were busy and Sorokin had joined them. After about ten minutes, one of the men replaced the receiver on the transmitter and they all stood up looking out toward the sub and waving. Within seconds, water and ice foamed up over the hull and the black shape slowly sank back beneath the waters of the bay, the conning tower and its antenna disappearing below the surface.

McClure looked at his watch… 09:42 AM. The whole

exchange had taken less than fifteen minutes. He watched as the men zipped up the white cover on the transmitter and began strapping it onto the bike. He took the hint and wiggled backward down the hill. With no idea where the second group was heading, if he wanted to get back in time to get their general direction and be able to track them, he'd better beat feet now. He quickly strapped on his snowshoes, picked up the sticks and rifle, and took off out from the spruce trees and back north. The trek would go faster if he didn't weave through the trees, so he stayed out on the open snow. His lunch of jerky and crackers would have to be consumed in transit.

The trip down from the Russians' camp and then back up again after only a fifteen-minute respite exhausted him. His calves and thighs ached, and his breathing grew labored. The sun had warmed the air into the 30s and while the temperature was pleasant, he opened the parka to keep from overheating. His water bottle would be empty soon too, but he knew he couldn't stop. From the sound of the engines, McClure guessed the snow bikes had already been at the camp a half hour by the time he got back to his viewing spot. It was 11:25 AM. He unstrapped the snowshoes, dropped the sticks, laid his rifle carefully on the shoes and crept up to the edge.

McClure yanked out the binoculars. All three bikes were in a row by the fire pit, so the second group hadn't left yet. Well, the bikes had to be refueled and perhaps some other maintenance was needed. The transmitter was sitting just inside the large tent. For a brief moment, he set the binoculars down and scooped up some snow with his left hand and stuffed it in the bottle.

Krasnoff strode from their tent over to Ivanov. "Alexi, you, me, and four others should be ready to go in ten minutes. We've lost some time and as it is, we'll probably be coming back in the dark."

"It was unavoidable, Major. We had to refuel the two bikes and then clean off some fouling." Ivanov replied blandly.

"I know. We just have to get going, that's all. Bring Sorokin and Mikulich, and two juniors. And make sure we have enough water and some dry food. What we have to do at the site will take some time. The other two will stay at the camp."

"Yes, sir." Ivanov walked back toward the supply tent.

Dave frowned. Ten minutes rest. This was going to be a very long day if he wanted to stay at least within earshot of the bikes. *Okay, but I know now where they came from, and if I can keep up with them on this leg, I might find out why they're here. Good,* he thought. He stuffed more snow in the bottle and set it in the sun, then pulled out the jerky and crackers. It was incredible how good this stuff tasted in the wilderness. Almost like a steak and mashed potatoes… well, not quite.

He smiled, squirmed backward and rolled over in the sun. It felt so good to just have some time to lie still. After just a few minutes passed, the sound of motors echoed up the ravine. He wolfed down the rest of the jerky and crackers, gulped some water and crawled back up to the edge. Krasnoff was on one bike, Ivanov on another, and Sorokin on a third, all revving their engines.

Krasnoff shouted above the roar, "I'll take the lead. You two follow. We're going north into the woods, out onto an ice field for a bit, then back into the woods. Pretty much due north and a little northeast. We'll come up on the edge of a pretty steep canyon, so be careful. You'll see a bridge. We'll talk again then. Are we ready?"

"*Uraaa!*" Ivanov shouted.

"*Uraaa! Uraaa!*" Everyone in the camp cheered back.

"Okay, let's…" His bike ready to lurch away, Major Krasnoff hesitated when he saw Junior Lt. Lipovsky running out of a tent and waving at him.

"Captain Ivanov, Major Krasnoff!" Lipovsky shouted at the top of his lungs.

"What, Lieutenant?" Ivanov's right hand sliced through the air in gesture to cut the engines.

"Sir, the satellite weather radio just alerted, the weather center in Anchorage."

"Yes?"

"A huge, fast-moving cold core system is heading south-southeast toward Valdez. It's big, sir. Barometric pressures are falling off precipitously, and they're warning of several feet of snow and turbulent winds up to 60 mph. Nome and Fairbanks are already in a maelstrom with visibility down to zero."

"Shit!" Ivanov shouted in consternation. "Major..."

"I know. Dammit. Alexi, it would be tough enough finding our way back in the dark. We'll have no chance at all in such a storm." He paused. "Well... damn, we must abort!"

"Men, we're aborting!" Ivanov repeated the command. "Maybe tomorrow or the day after, Major."

"Yes, no worry. We'll find the means."

Ivanov eyeballed the men gathered around them. "Men, in a few hours it's going to feel like we're back on the Russian steppes in the dead of winter! Let's make sure everything is stowed securely, all well tied down. Secure your tents also. Some of you go out and gather extra firewood, we may need it. Okay then, let's get to it. Then, maybe we will search for some vodka to keep us warm!"

"*Uraaa!*" The cheer rang out as the men hurried off in all directions.

"You brought vodka too, Alexi?" Krasnoff questioned, surprised.

"Half a case of Stolichnaya, Major. It's packed in an ammo case."

"Hah! You fox!" Krasnoff laughed. "You are a captain, but

you already think like a major! Well, let's split up and help the men prepare for this storm."

Ivanov nodded in reply as he turned his bike toward the tent, wondering if he had just been complimented by this skunk.

Dropping his binoculars, McClure grimaced. "Well, hell, that's just peachy. A blizzard yet," he muttered, pushing back from under the spruce and gearing up. He slung the rifle over his shoulder and plodded out from the trees. In just a half hour, he found himself struggling to focus on the rhythm of his pace, and even more so on his already labored breathing. He looked up as he trudged north. High cirrus ice clouds were sweeping down from the northwest, turning the sun into a faint, ghostly ball in the sky.

"Well, Dave, you know, at this very moment you could be back on your couch in Colorado Springs sipping a beer, munching nachos and watching the Denver Broncos pound Kansas City. But hey, why do something so mundane? No way… just think of the adventure you're experiencing here, Dave, the very thrill of it all!" He gasped aloud. "Idiot."

The tops of the spruce trees were beginning to move, waving back and forth as the northwesterly winds increased. The skies to the north were dark and menacing. McClure swung his wrist up and glanced at his watch… 12:30 PM. Lunchtime, no time to stop now though. He was dead-ass tired, his whole body ached. But he battled to pick up the pace, knowing full well that he had to get to the campsite before the brunt of the storm hit. It could very well be a question of survival.

CHAPTER 17

Alaska

A vortex of winds moaning like a banshee engulfed the iced meadows to McClure's right and the chilled forest to his left. The clicking and clacking of birch and aspen branches rattled in the air like snare drums and pounded against his ears. The snow hadn't yet begun, but a ground blizzard of wind swirled around him, scattering ice and snow in every direction. He noticed that snow, driven by the blustery winds was now drifting into his tracks on the path ahead. Not good. The last thing he needed now was for his trail to disappear.

Dave fought to increase his pace, but had simply no energy left in him. He was starting to sweat under the parka and knew very well that wet plus cold equaled hypothermia. But he couldn't stop now even though pain stabbed his lungs with every breath from sucking in the polar air. He looked at his watch... 1:45 PM. He had to be close.

The goggles protected his eyes well and added clarity to the surroundings, but the windborne ice stung his face. He'd always found fields of snow to be beautiful to behold, but having tons of it blowing right at you lessened the appeal. Visibility was closing in. He glanced upward to see dark clouds racing in from the northwest, driving the storm's first flurries. Huge flakes were soon swarming in the air around him like a horde of angry white bees.

Unexpectedly, McClure saw the familiar break in the tree line to his left. *Oh! Praise God!* He angled his steps toward the woods, feeling a swell of renewal, and slogged feverishly through the snow, relieved that he was so near the campsite. As he entered the trees, the piercing force of the wind dropped precipitously, blocked by the barrier of black spruce, aspen and birch. At the same time, he noticed an icy crust on the surface snow that had begun to crunch under his snowshoes. Without the violent force of the wind, the ground cover became visibly clearer. Dried branches and dead grass poked out from the drifts and waved gently to and fro.

Other than the persistent rattling sounds of the branches overhead, the forest seemed to have come to a standstill, the woods paralyzed by the frigid onslaught. No sign of life whatsoever. The snowy owl, the birds and the ptarmigans were all absent, probably huddled down against the fiercely advancing storm. Peering ahead, he spotted the tent in the small clearing not twenty yards away. A wave of relief washed over him. He'd made it.

As Dave stepped into the small camp, his mind was already buzzing. What should he do first? He was desperately fatigued, but he was also thirsty and starved. Water first. He hunched down and unzipped the tent flap to check inside; it was dry as a bone. Good. The tent's entrance was facing away from the gusty northwesterly winds, also good. He eased the rifle off his shoulder and laid it in the tent next to the sleeping bag. Squatting, he scooped some fresh snow, stuffed it in his bottle and slipped it under his parka into the belt holder. It would melt in minutes.

Standing up, he looked at the fire pit. Clearing it of the new fallen snow consumed several minutes, after which he paused for several gulps of fresh water to quench his parched throat. With brush and branches from the stack by the tent and a few shavings from his magnesium block, McClure had a nice, blistering fire in no time. He drew his face and hands close to it, enjoying the delightful feeling of its radiating warmth. The fire would let him cook an ample late afternoon dinner before the full impact of the blizzard descended on the campsite. It might be his last good meal for a while. Hell, he might finally grill that moose tenderloin steak! All things considered, the situation was looking better by the minute. He'd had a close call.

Dave pivoted around, his eyes surveying the campsite as the wind dwindled and snow began falling silently and steadily in flakes as big as goose feathers. The encircling woods were beginning to look like a scene from Dr. Zhivago. A quiet pageant of winter splendor. *Anne, Michael, this really is so gorgeous. I wish you could see this. Then again, maybe you can.* The thought drifted through his mind. He could almost sense their presence, envisioning their faces smiling at him. He grinned cheerfully as he watched the snowfall enfold the trees and ground. It was wonderful to think of Anne and Michael, and to do so without the abysmal depression he'd become so accustomed to.

It wasn't possible to estimate how long before the center of the blizzard arrived, which would force him into the tent. The low pressure system seemed to have slowed. So, what else did he need to do before light completely vanished from the forest? No need for a perimeter alarm tonight. Not a creature moved through the forest. He turned around, considering the tent. With this wind, snow would soon pile up behind and over it. How about a sturdy lean-to? He could build a relatively narrow shelter just behind the tent to shield and protect the tent's space, diminishing the effects of the wind and catching any drifting snow. Right, that would be the next chore.

Dave walked into the woods and with the serrated side of his SOG knife blade, began cutting the straightest trunks he could find. An abundance of sapling trees grew nearby, just the size he was seeking. Six more substantial trunks, two with a branching notch at their tops would serve as front support stakes, with a crossbar between them and three longer stakes leaning up against the crossbar to hold the smaller branches and spruce boughs. He found them all in short order. Dave used the camp shovel to set the support stakes and snipped some lengths of the steel tension cable and the nylon rope to lash the frame in place. Finally, he interlaced the branches and spruce boughs on top of the lean-to.

He stood back and admired his work. The whole job had taken a little over an hour. "Not too damn shabby, McClure... nice job. Feel free to reward yourself with dinner and coffee." He spoke out loud, chuckling.

Dave raised his water bottle and drank the whole thing down. His body was invigorated with the hydration. After stuffing more snow in the bottle and the coffee pot, he stepped out the distance to where he buried the moose meat. He retrieved it and cut off a generous slice for a grilled steak, reburying the remaining chunk for another day.

Sticking a stake in the ground on either side of the fire, he shaved the bark off a third and whittled it to a point as a makeshift spit. He sat in front of his tent on a small fluff of spruce boughs, watching the steak grill and the water come to a boil, his mouth watering in anticipation. Looking up, he saw the tree branches whipping back and forth vigorously. The wind was now gusting through the campsite, finally penetrating the thick woods.

By the time the steak was done, the storm was really upon him. The white stuff was beginning to pile up. He chowed down on the steak, savoring every bite. Robert was right, it was wonderful... juicy, simply delectable. Then the coffee and a couple crackers. He cleaned up the pot and utensils in the snow until they were sparkly clean, reached in the tent and stuffed them into his bag. Squinting over the top of the tent and into the flurries, it was apparent that the lean-to was already performing its task well. He was as ready as he was going to be. And what a day it had been. He started to douse the fire, but realized that the storm and its wind and snow would take care of that. No worries there.

It was time to hunker down. He had the sterno stove and two cans of the stuff left, in case he had to have breakfast in the tent. That would work fine if he opened the tent flap a bit for ventilation. He crawled into the tent and zipped the tent flap almost to the bottom. After shedding his boots and parka, he squirmed into the ice-cold sleeping bag, gasping at first with the sudden chill. If nature called and he had to go out into the woods to take a piss, no doubt his stream would freeze to ice cubes before it hit the ground. That thought only made him colder, so he pushed it from his mind, tossing and turning to generate some warmth. In minutes, however, he could feel the bag radiating back his body heat. He turned over on his back, staring into the dark shadows of the tent above him. The fabric rippled with the wind howling outside. He could feel a slight draft even though the tent flap was closed.

Though his eyes were open, he could see nothing but blackness. Then, his sight blurring with a heightened sense of fatigue, the vision came out of nowhere, blasting its way into his consciousness as vividly as the storm roaring outside the tent. He remembered every detail with stark clarity. He sat at a temporary desk in the Cleveland Field Office on 9th Street when the call came. The front desk put it through to his desk.

"Special Agent David McClure...?" The duty officer at the Washington DC Ops Center asked tentatively.

"Yes? This is McClure."

"Special Agent McClure, sir, you need to get home immediately. We just received an alert from the Colorado State Police.

There's been an accident... your wife and son... on Interstate 70 near Vail. A bad accident. Sir, you need to get home."

"An accident? How bad? Are they okay?" The hair on his arms began to rise.

Silence on the other end.

"Tell me, dammit! Are my wife and son okay?"

"No, sir. I'm sorry." The duty officer paused, collecting himself. "Sir, please catch a plane back to Colorado Springs soonest, tonight if you can."

"Hold on, don't you hang up on me!" McClure's voice was now desperate. "Both of them?"

"Yes, sir. I'm very sorry, sir. That's all I know. I'm sorry. God bless." The phone clicked off.

McClure felt the nausea rising in his chest. He yanked the wastebasket out from under the desk, fell on his knees and puked his guts out. Other agents in the office clustered around him, but could offer no solace. They didn't know Dave McClure's world had just collapsed. A week later, the double funeral was held in Colorado Springs. It was a cold, cloudy November day. A cutting wind blew across the grounds of Memorial Cemetery. He remembered the faces, people offering sympathy.

The grief was a vise, crushing his mind and spirit. He had nowhere to let go of it, no place to run away from it, no way to pretend it hadn't happened. Week after week, he found himself visiting the graves, gaping at the dead, frozen grass above the two people in the world he loved most. On a freezing day in January wearing nothing but a topcoat in the blowing snow, he stood saying good-bye yet again and trying to make some sense of the tragedy. But of course, there was no sense to it.

Still staring into the blackness, Dave blinked away the tears, remembering Pete Novak's words: "You don't get away from it; you learn to deal with it. Dave, the way to do that will come to you in time." Dave had been feeling for a while that he'd found his way out here in the remote icebox of the Alaskan winter. Like Jack Stroud had learned to do. But unlike Stroud, it was possible he would survive this winter sojourn. He could go back... no, would have to go back. His love for his wife and son was stronger and more powerful than his grief. He scrunched his eyes closed, turned on his side and huddled into the soft fabric of the bag as the blizzard swallowed up the wilderness.

Miles to the south, the Russians were discovering they'd made a terrible mistake by clearing out the birch and aspen trees in the ravine. Now, no barrier existed to protect the camp; nothing impeded the wailing winds that raged down through the ravine's north entrance, propelling snow, ice, brush and branch debris and thrusting it south through their camp. The snow men struggled vainly to keep the fire going in such violent wind, eventually giving up, dousing it, and then covering the pit to keep it from filling with snow. The tents were flapping furiously in the blinding wind, and waist-deep drifts of snow rapidly accumulated around them.

All the team members huddled in their leaders' tent, talking, joking and drinking vodka. Vassily and Alexi also surprised them with crackers and six tins of beluga sturgeon caviar. Alexi, however, was beginning to think the celebration was going too far. After all, they were on a mission of great importance, according to the yo-yo of a major. Alexi had stopped after just two drinks and was now watching his men as the booze numbed their minds and bodies to the blizzard roaring just outside.

Alexi didn't like it. Periodically, he sent one of the junior lieutenants out to check on the bikes and transmitter, the things so critical to their mission. But, some of these men were getting so shit-faced the captain wondered if they could even stumble back to their tents to sleep it off. Perhaps he should have held off on revealing all the vodka.

Vassily saw it differently. He wanted to give them all the opportunity to cut loose, hence the vodka and caviar. In the next couple days, he would ask a lot of these men. He was going to ask them to kill, and kill relative innocents, and then do something he was certain they'd never even dreamed of. Even Captain Ivanov didn't know their mission details either. No, he would wait until they were at the site to tell this pain-in-the-ass captain and his fabulous Spetsnaz team what their orders were. Some of them could die, and he would ensure that they clearly understood this was for Russia and the Motherland. The difference was that Vassily himself knew he and his Russian mafia brother, Leonide, would be handsomely paid. They would retire together and disappear into the Pacific sun, so far out they would probably never hear the Russian language again. *So, drink up men. The blizzard will soon end. And your core mission will begin.*

CHAPTER 18

Israel

Taavi Perutz drummed his fingers along the edge of the fortress of mahogany he called a desk. His eyes swept from face to face, General Goren to Raphael Mahler to Meira Dantzig, then glanced out the window and back to his desk. The walls in the office were filled with photos of Taavi and famous personages: one had him sitting on the top of an armored personnel carrier in the desert with Moshe Dayan. Another had him on a couch sipping tea with Golda Meir. And yet another had him on stage in the Dome at HQ CIA next to George Bush, the elder.

"Ari, thanks for getting me engaged, and thanks for the briefing. If you are correct, then this is truly very serious."

"Sir, I wouldn't have called if I didn't think we had enough indicators at this point to involve you."

"Of course, Ari, I know that. And, I meant 'thanks' sincerely. All of us here are in the business of intelligence. The intelligence we gather must be used wisely in ways that protect our nation as well as our sources. You know as well as I do that very little comes our way in nice, neat packages of actionable intelligence. On the contrary, we have to sit down together, examine indicators from all reliable sources, and then force ourselves to think outside the box to put the pieces of the puzzle together as best we can. Yes?"

"Yes, sir, I fully agree."

Perutz turned his head toward Raphael and Meira. "And so, thank you both for doing the digging, the legwork, and for putting together what pieces you had. I have to say, however, that I hope with all my heart that you are wrong, as I am sure you do. But, there is too much on the table right now to just walk away from it. I agree that we should involve the Americans, at least make them aware of our suspicions, that is, without compromising our sources or our collections methodology. Agreed?'

They all nodded in reply.

"Truth is, this has always been my worst fear. On one or two occasions, perhaps going to sleep after too much wine, I've even had nightmares about it." He paused, his eyes looking far away, shaking his head weakly. "Well, let's see, it's a little after 11:00 AM here, so it's only about 4:00 AM at Langley Center. I doubt that the deputy director, my friend Peter Novak, is at his desk this early. Can all of you come back here at 4:00 PM? It will be 09:00 AM on the U.S east coast then. We'll go into the secure room and call."

All of them nodded again.

Perutz pressed a button on his desk phone and. "Sharon?"

"Yes, sir?"

"Please reserve the secure room for me at 4:00 PM for a half hour. Call Nathan and ask him to be there. I want him to make the secure-call connection for me to Langley Center in Northern Virginia."

"Yes, sir."

"Thanks, Sharon." He pressed the intercom off. "Well, we have a few hours. Go take a break, collect your thoughts. Be back here at quarter 'til four and we'll see if we can get Mr. Novak on the phone, see what he has to say. Thank you all." Perutz rose from his seat, dismissing them.

"Thanks, guys, see you in a bit." Ari patted Raphael and

Meira on the back as they left Perutz's office.

"Meira, how about a sandwich out in the courtyard?" Raphael asked as they approached the elevators.

"Sure, sounds good."

Raphael bought his traditional chicken-salad sandwich and a Coke, Meira a turkey and cheese with iced tea. They picked a table in the shade.

"Pretty ugly tree, isn't it, Meira? Even the leaves are ugly, kind of hairy and something underneath." He stared up at the branches. The trunk was fat, four feet thick, and its branches were gnarled and twisted.

"Well, Rafi, if it gives shade, who cares? It's a Mount Tabor oak. The tops of their leaves are shiny, and the bottoms are dull and have small hairs. Those small hairs are what repels insects. Some of them are hundreds of years old. But I agree with you, I like the pine trees better."

"Gees. You a botanist too?"

"No, but my father was. Flowers, plants of all kinds, trees... he loved them all. I guess some of it rubbed off on me. He was very excited when talk began years ago of reforesting the Negev. And now they're actually doing something with it, bringing in water and growing plants. Scientists from all over the world are working on projects out in the desert. I wish he could have lived to see it."

"I'm sorry. How did your father die, heart attack, cancer?"

"None of those things. In August 2003, he and an old friend made the mistake of walking down to the corner store to get a soda. He loved root beer. He was in the Shmuel HaNavi neighborhood in North Jerusalem, oh, just for a short visit. I remember that day vividly. I was at the Hebrew University campus checking things out. I was 17... going to matriculate there in September."

"If I may ask, what was your major?"

"Computer science. Anyway, my cell phone rang. It was my mother. She was hysterical. Police were at the house. Some 29-year-old preacher from Hebron disguised as a Haredi Jew had gotten on a bus stopped in Shmuel HaNavi, came in through the rear door, and detonated a bomb. It had been purposely spiked with ball-bearings to increase injuries. It killed 23 people, seven of them children, and one of them a young pregnant woman."

"Meira, I am so sorry. But… your father wasn't on the bus, was he?"

"No, but he would have suffered less if he had been. In the blast, a metal-framed window blew out from the bus like a Frisbee and hit him in the stomach, ripping into his colon. He survived the four hours of surgery that night, but endured excruciating pain from intestinal problems for more than a year as he withered away and finally died."

Her eyes were wet, but she continued. "You know, his name is not carried on the casualty list of that bombing. But there is no question, Rafi, that bomb killed him. Hamas claimed credit. Months after the attack, IDF special forces raided a safe house in Hebron and killed the other Hamas members responsible for the bomb. I was happy to hear that, of course, but it didn't help the grief my mother and I agonized with after his death. Ezra was a good father, a good man. I still miss him."

"Meira, I had no idea. I guess I really know so little about your life. Please know, I am so sorry for you and your mother. She's still alive?"

"Thanks, Rafi. Oh yes, Mom is still alive… if you can call it that. She sits in a chair all day looking out a window to the little garden they have, I mean she has, in back of her house. She won't move to an assisted living facility. I arranged for someone to come in and prepare her meals for her, which helps a little. You know, Rafi, all of that drove me here."

"I'm sorry, what?"

"Why I joined Mossad. I was young, had just graduated from the university, but I was so filled with anger, I wanted to kill as many of them as I could. Now, after shooting that young man in Jerusalem, though, I don't think I have that anger anymore. Just depression, and wondering why things have to be as they are, so much hate."

"I think that's a natural reaction to what happened to you, Meira. General Goren told me he considered you a warrior and I agree. He also said that you never know what someone has in them, what they're capable of, until that critical moment comes. You showed what you're made of, Meira. You are courageous. This will all pass, believe me."

"You really think so?"

"Yes, you just need time. They say 'time heals all wounds.' Some things we never forget, but time takes away the burning ache. This too will pass."

"Rafi, I love talking to you." She leaned over the table and pecked him on the nose.

"Meira, please…" Rafi protested, but he was smiling.

"Let's just sit here for a while. It's such a nice day. The trees, the flowers, the people… they're all so beautiful, each in their own way.

"You sound like a romantic rabbi, but okay, just for a while." He sipped his cola and settled into his seat. Time passed with little or no conversation, both of them inundated by their own thoughts. People were leaving the courtyard. Finally, Rafi stirred.

"Meira."

"Yes?"

"I'm going up to my office, take care of some work I still have hanging and clear out some emails. I'll meet you at Perutz's office at 3:45 PM. All right?"

"Sure, go ahead. I think I'll linger here a bit longer, then go to Kirschner's office to let Ethan know what I'm doing, what's happening. Take care."

"See you." Raphael strolled across the courtyard and disappeared through the double doors.

CHAPTER 19

Israel

After two and a half hours of waiting, Raphael and a still nervous Meira were standing in Perutz's outer office when General Goren approached, a solemn look on his face.

"Ready for this?" he asked.

"I think so, sir. It's so serious. I think we're doing the right thing," Raphael answered promptly.

"We are. Come, let's go in."

Perutz was at his desk. When he saw them, he stood and put his suit jacket on. "That time already?"

"Yes, sir. By the way, nice suit," the general smiled.

"And what would you know about suits, Ari? You're in uniform most of the time." He chuckled. "But thanks. It's an Armani, bought it in Rome. I just had to have one. And what about the Zegna tie, eh?" He turned half-profile, posing and smiling. "This necktie set me back more than the shirt."

"Zegna? Never even heard of it."

"See, that's what I mean, what the hell would you know about suits! Military operations? Yes. Suits? No." Perutz laughed out loud, came around the desk and clapped Ari's shoulder. "Come on, my friends, let's go get this done. Nathan said he'd meet us outside the vault."

Perutz and Goren led the way through the outer office and

turned right down the hall. Raphael and Meira followed, exchanging nervous glances. As they rounded a corner to their left, they saw Nathan Kissner, Perutz's communications director, waiting for them by the vault door, his hands in his pockets. He straightened up at the sight of Perutz.

"Shall we go in, sir?"

"Yes, Nathan. Thanks for being here."

After sweeping their proximity access cards and entering their pins, all five walked into the empty vault. They took seats around a large desk, with Nathan on a swivel chair with rollers in the center.

"Sharon said this call was to Peter Novak, Deputy Director of Operations at Langley Center."

"That's correct," Perutz answered.

Nathan swiveled his chair and pressed some buttons on the display. After a few seconds, a green light in the middle lit up, and he entered a series of numbers then flicked the speaker switch to on. The line rang just once, and the ring stopped with a click. Nathan slid his chair out of the way and Perutz scooted his chair closer.

The speaker phone crackled. "Hello. This is Cathy Harding."

"Hello, Ms. Harding. This is Taavi Perutz, Tel Aviv. If I may, is Mr. Novak available? I have some important information to pass on to him."

"Mr. Perutz, good morning, sir. Yes, Mr. Novak's in. I'll put you through."

"Ah, Ms. Harding, Mr. Novak will have to go secure on this call. Can it be transferred?"

"Yes, sir, that's no problem. He can do that from his desk. I'll put you through."

Pete Novak was sitting at his desk, sipping coffee and reading the morning's regional intelligence messages from the station chiefs. A lot was going on around the world. His phone rang, the gray phone.

"Yes, Cathy. Secure call?"

"It's Taavi Perutz, Tel Aviv."

"Oh. Please, put him through." He dropped the intelligence reports on his desk and heard a series of clicks as the line went secure.

"Taavi? Pete Novak here."

"Pete, good to hear your voice. How's the weather?"

"Ah, cloudy, windy and in the high 40s. You?"

"We have about 62 degrees right now and sunny."

"Thanks a lot." Pete laughed, and Taavi joined in. "Taavi, what's up, what can I do for you?"

"I know, that's usually the way it is. This time though, it's what we want to give to you, Pete. Do you have a secure room close by that you can get to without too much trouble?"

"No need, Taavi. This whole building's a secure room. I've asked you several times to get away from your desk and make a trip here. I'd like to show you around, let you see what we've got now. We could even go to dinner down in Old Town Alexandria. You'd love it."

"Pete, how…? You can go secure from your desk?"

"Taavi, this whole building, the whole facility is encased in a metal-glass alloy cube. We run a current, low voltage, through it. Not an electron escapes. No Emanations Security, ah EMSEC, concerns here. On the outside, it looks just like any other glass and steel building. So, yes, I am on a secure line with you right now. You really should come visit. I'll make it worth your while."

"About the building, is that classified?"

"No, it's not classified. We just don't talk about it much."

"Well, I really would like to see that, Pete. Thanks for the invite, I'll work on it. So, to business… by the way, I have General Ariel Goren with me and two of our agents, Raphael Mahler and Meira Dantzig."

"Ari, hello! How are you?"

"Just fine, Pete, thanks. Raphael and Meira are here because they developed the information that we want to pass on to you. Mr. Perutz fully agrees," Goren replied.

"Sounds serious, Ari," Novak said, his tone no longer jovial.

Perutz responded, "It is, Pete. Very serious. I wouldn't call you like this if..."

Novak interrupted him. "I know that, Taavi. What is it?"

"Pete, we believe that there is a plan by the Russian mafia and Vevak, the Iranian intelligence service, to steal a Minuteman III MIRV warhead. We don't have a clue as to how or where. But all the indicators, the intelligence we've pieced together, point to it."

"Steal an operational ICBM nuclear warhead? What the... that's it? Anything else?"

"No. But everything we've accumulated suggests it."

"And the Iranian government just announced they were opening up the whole damn country to inspection and giving up further development of nuclear weapons-grade material."

"We don't believe it."

"Well, truth is, we don't either."

"Pete, this actually fits in: an Iranian media announcement part of the mask for a strategic deception operation. No one in the world but us and maybe the Brits would expect this kind of thing now."

"Right. Damn. I... I'll need to run down the director and brief him. So, you don't have any information concerning how such an attempt would be made? That makes it tough. You know, we only have three operational ICBM bases and they're all out in the middle of nowhere in our northern states, Montana, North Dakota and Wyoming. And secure as hell. Nobody can get anywhere close to those bases without our knowing about it."

"Yes, we know. Actually, the team has focused on Alaska, but I can't give you anything actionable in that regard. It's a best guess. I hesitate to give you a recommendation, Pete, but at this point you may want to keep this in intelligence channels. You know, rather than energize the whole defense system. Perhaps just declare a higher Threat Condition at just those three bases. Notch the THREATCON up a level."

"That's the director's call. He may delegate this to me, though, and we'll talk to our top commanders at the Air Force Major Commands controlling those missile fields. Alaska, huh? Interesting. We'll keep that in mind. Taavi, I like your suggestion of keeping it in intelligence channels except for the Air Force senior command elements and the security forces at those three bases. I'll let you know how we're handling it. Believe me, I understand your concern. Your stake is the highest in this."

"One last thing, Pete."

"Yes?"

"I'd like to offer you the loan of these two agents, Raphael and Meira. They've been the lead on this for us. They could help."

"Another good idea, Taavi. Thanks for the offer. Okay, get them over here soonest. Let us know their flights and arrival times at Dulles and we'll pick them up. Oh, and the information I'll need for their non-escort badges, clearances, etc. I think we'll wait until they arrive to start making calls, they may be able to assist in that. Don't worry, Taavi, we'll take good care of them. Oh yes, have them to bring their personal weapons, okay, Raphael, Meira?"

"Yes, sir."

"If our people aren't at the gate when you arrive at Dulles, ask for directions to the Government hangar. They'll understand. Your luggage will be hand-carried to our hangar. We'll catch up. Actually, I'll try to be there to meet you myself. You work for a good man over there. Oh, and there won't be any customs inspections."

"Sir, thanks." Raphael answered.

"Pete, great. Ari and I both thank you for listening."

"My job, Taavi. We're all brothers in this business. We have to be. Goodbye, Taavi, Ari. Raphael, Meira, see you soon. Novak out." The line went dead.

As Taavi Perutz switched off the speakers, he looked around the desk. "Good. Well, we've done what we had to do. Thank you all. Things are going to get a little crazy on Pete Novak's end. I guess they have to. Raphael, Meira... go home and pack. You've got the rest of the afternoon and the evening. I'll ask Sharon to book your flights - a late morning flight tomorrow through Zurich. That's a nice flight. You'll go first class... I want you as rested as possible when you land at Dulles. You should be at Washington Dulles by 8:00 PM. So, get going and don't waste any more time here today. Good luck and God bless you. Shalom."

They shook hands all around, then scattered once outside the vault. Raphael looked appraisingly at Meira as they headed down the hall.

"Nervous?"

"Yes, nervous. But Raphael, I dread the possibilities in all this. I mean, what horrible devastation it might represent. But then also, yes, I'm excited too. We're going to America!"

"A big yes to that... this will be an awesome trip. However, you're right in that we can't lose sight of 'why' we're going. This is deadly serious business for us. Do you need to go by your office? I can wait."

"No, I have everything I need is in my bag at home... gun, extra mags and ammo too."

"Good... well, your car's gonna have to hang out in Jerusalem for a while. I'll drive you home, and I'll pick you up tomorrow at 7:00 AM. We won't have to rush, but we can't lag either," Raphael said.

"Crap, I forgot all about the car. What the hell, it'll still be there when we get back, whenever that is."

The ride to the apartment was brisk. Buildings, fields, everything seemed to blur by. Raphael swung the Passat into a parking place up front.

"Just walk me up?" Meira asked.

"All right."

They walked arm in arm into the elevator and then down the corridor on her floor. Standing outside the apartment door, the two looked silently at each other for a brief moment. Meira put the key in, clicked the lock, pushed open the door, and turned to Raphael.

"Come in, Rafi."

"Meira, no..."

"No walls," Meira whispered, sliding her hands up his chest and around the back of his neck, pulling him toward her. She arched her neck and kissed him full on the mouth. "Please," she murmured, kissing him again, as hard as she could and drew close, pressing against him.

Raphael looked into her eyes and shook his head slowly, a small smile forming. His arms skimmed around her waist and his lips met hers, then slipped lower, sliding along her neck.

"Ohhh..." She pulled him closer.

As his hands tightened around her waist and slid down her hips, Meira inched back into her apartment, pulling Rafi with her. They kissed again, their arms intertwining. She reached out and shut the door.

CHAPTER 20

Alaska

The silence was deafening. The abundance of snow acted like an acoustic sponge that sucked up all sound in the forest. But eventually, soon a few chirps outside the tent defied the stillness. McClure wiggled out of the bag and peered out through the bottom of the tent flap. He could make out movement in front of the tent. He listened to the twittering for a moment... *the chickadees and sparrows again*, he thought. They were pecking at the dry grasses that had still somehow managed to breach the snow's surface in the lee of the tent. He could just make them out against the backdrop of snow.

He looked down at his watch... 07:10 AM. It was dark but growing light to the east. Drifts of snow blocked his view to both left and right, so the lean-to had done its job preserving, the tent space and the fire pit in forming a narrow valley relatively free of the white stuff. He unzipped the flap slightly, inched forward, turned on his back and looked up at a sky filled with stars slowly disappearing east to west from the early morning grayness. The storm front had passed through and the wind was gone.

McClure wanted to be there when the Russian snow men climbed out of their tents which might be difficult this morning even though the snowshoes would work to his benefit. Digging through his pack, he discovered his provisions were extremely low. He'd been ignoring the obvious: the pack food wouldn't last forever. He still had two packets of chicken soup, instant coffee, a couple pieces of jerky and eight crackers. Oh yeah, he still had at least one good moose tenderloin steak as well.

McClure remembered Ed Weiss mentioning that grilled rabbit wasn't bad at all. Like a lot of things, it tasted like chicken. And there were huge populations of rabbits out here. He couldn't count on Robert and John Ewan showing up again and donating some additional game, but there was a remote chance. Before getting underway, he needed to find a couple good spots to set out some loop snares.

His parka, boots and gloves donned, he crawled out of the tent and walked over to the fire pit. The drifts around the campsite were a little over three feet deep, but he could see branches and twigs sticking out where his stack of kindling was. After brushing snow out of the pit and stuffing some twigs under a teepee of branches, he tossed a windproof flaming match into shavings from the magnesium block. The fire took immediately, rising in a small blaze through the layered branches. When he placed some larger branches on top, the fire began to sizzle and pop. The radiating heat felt delightful.

He observed that the cooking spit survived the high winds, so why not have another grilled moose tenderloin steak for breakfast to provide maximum energy for the day's activity? It sounded great, however the storage spot where he'd dug out for the meat was now buried under at least three feet of snow. Retrieving it would take some effort. Using his pack shovel, McClure took on the task of digging out a path to storage site, removing a snow drift, and finally retrieving the rest of the meat. He would burn the wrapping material in the fire. At the fire pit, he added two crackers and a cup of coffee to the grilled steak and devoured his breakfast, again reveling in the wilderness experience. Cleaning up took just a few minutes, and he was ready to set the snares.

With the high tension aircraft cable, a couple stakes and crackers with a little dry chicken soup sprinkled on as bait, McClure patiently went about the chore of setting three loop snares. He found more than enough sturdy saplings to bend over as triggers for the snares. All were carefully nestled in small stands of birch and aspen about twenty-five yards from the campsite. With any luck, he'd be dining on grilled rabbit this evening, washed down with coffee. In his parka, he stuffed the last of the crackers and jerky, and hooked the water bottle on his belt. He checked both pistol and rifle and dropped an extra box of ammo for each in his pockets. McClure zipped the tent flap, doused the fire, strapped on his snowshoes, and set off from the campsite with the rifle over a shoulder at 08:30 AM.

Not knowing exactly where the Russians were headed, it made sense to him to rejoin the group at their camp and then parallel them from the east. He would need to maintain a safe distance between them to stay out of sight. He tramped out of the tree line and headed south just as the sun was rising over the adjacent meadow blanketed with new snow. White, white and more white. The brilliance of the sun on the seemingly never-ending field of snow was dazzling. He yanked down his goggles. The snow lay in piles of waist high drifts. Maneuvering around them slowed him down, but he also found large patches of ground between the drifts blown bone dry by gale force winds.

After an hour of trudging southward, he saw a rough path through the snow crossing the meadow and disappearing into the woods. He hunched down and peered into the bottom of the snowy path. Bear tracks. Bear tracks soft and crumbly on their edges. They had to have been made sometime late in the night. He grimaced, having no desire to meet up with either

Bruno or Brutus ever again.

McClure soon entered the line of black spruce that concealed his observation point and trudged in among them. As before, he slipped off the snowshoes and rifle in the small open area, and wiggled up to the edge of the ravine sheltered by the boughs of a snow covered spruce. With relief, he realized he arrived with time to spare. The camp was bustling with activity. Some men were still digging out the tents while others struggled to retrieve the large tent, which had apparently torn and was half buried, along with most of their supplies.

The snow bikes were lined up by the fire pit, one man brushing the snow off and another kneeling to attach short skis under the front wheels. The camp had been hit hard with probably three feet of snow all the way down the ravine. *That's what you get when you cut down all the trees around you and remove your protective barrier, dumb shits,* McClure thought. They should've been smarter than that. Then again, who would expect a storm of that proportion in early November? *Well, fellas, Scouts' Motto: Be Prepared.*

Major Krasnoff and Captain Ivanov emerged from their tent and tramped down to the fire pit. Krasnoff waved his hands over his head and called, "Men, come gather around." The men dropped what they were doing and clustered around him.

"I know very well that there's still much to do here, but we need to follow our priorities. For the second phase of our mission, Captain Ivanov, Senior Lieutenants Sorokin and Mikulich, Junior Lieutenants Sokolov and Bardinoff and I will take the three snow bikes north to the site. We'll be gone at least five hours. Lt. Lipovsky, get your snowshoes and use the sled to take the transmitter down to the shore. Raise the sub and tell the captain about the blizzard, the delay it caused, and that we're heading north to the site for phase 2. Lt. Zaretsky, stay with the camp. Be alert. Let's get started. We leave in ten minutes. Are we ready?"

"Uraaa!" Captain Ivanov bellowed.

"Uraaa! Uraaa!" they cheered back.

It was time for McClure to get going as well. He knew he'd never be able to stay up with the Russians even with a slight head start, but at least he could try to stay within earshot of their motors. If they got too far ahead, he'd also have their tracks to follow as well. He quickly donned his gear and trudged out of the woods and into the open meadow. Without the goggles, the vivid radiance of the snow field would have blinded him in minutes. An incredible glare washed over the landscape.

He steadily tromped northward just outside the tree line, weaving his way around high drifts and nearly sailing across the bare patches in between. He made good time, the snow was dense and the snowshoes kept him from sinking more than six inches. In just over twenty minutes, his ears caught the loud whine of the snow bikes' motors due west of him. The Russians had climbed out of the ravine and were heading north through the forest.

The ice fields were to the east, so McClure intended to travel along them, keeping a good deal east of the snow men which meant that at some point, he might cross their tracks. If Krasnoff was correct in his time estimate, there was a good chance they'd be returning in near darkness. So, perhaps they wouldn't notice his tracks. Just in case, however, he should probably plan on moving his campsite.

In not quite two hours, he looked west and saw the break in the tree line that led to his camp. The Russians must have been traveling well west of it. Although he would be crossing their tracks within the hour, he was well out of his comfort zone, going this far to the north. He had a compass but never expected to have to use it. He was no wizard at land navigation, especially with no topographical maps and most especially not at night.

Another hour of snowshoeing and he could sense the land rising as he continued north. He was back in a forest of birches and had crossed the Russians' tracks, but could no longer hear the snow bikes. Maybe he was getting close. As he continued north, the trees thinned and he noticed a large gulch to his right, to the east. He stopped to take a look. It appeared to be more like a canyon, a deep one, because he couldn't see the bottom. The canyon curved gradually to the northeast, and as his eyes followed the slow turn of the canyon walls, he noticed a bridge spanning the canyon maybe a mile or more away. He abruptly dropped to the snow. The Russians were at the bridge.

Men in white parkas and pants were out on the span, but even with binoculars, McClure was still too far away to see what they were doing. He stepped back into the woods and moved further north for a better view. After another 500 yards, he ventured out toward the canyon again, shed his snowshoes and squirmed up to the edge, his rifle alongside him. Through the binoculars, he could see that they had moved over to the eastern side of the bridge. Krasnoff and Ivanov were both working with two lieutenants on the rails.

The track was already cleared of snow drifts, one of those giant rail-driven rotary snow blowers must have come through, maybe one of the Union Pacific's. That also meant that this stretch of track and the bridge were relatively high priority. McClure knew the Union Pacific Railroad ran up through Canada… how in the hell did he know that? He wracked his brain but couldn't come up with an answer. What was it? He couldn't remember. Something about this track was tucked away in some musty corner of his mind.

He thought he might actually get a better view through the rifle scope, so he slipped the rifle off his shoulder, popped off the end caps and focused the parallax on Krasnoff's head. It looked like the major's team was using a track ratchet wrench to test their ability to remove bolts from the rails. Ivanov's team was pulling out and replacing spikes with a track spike lifter. Then Krasnoff gestured and everyone gathered at the east end of the bridge. Krasnoff was obviously talking to the team, but McClure was too far away to hear anything, though he could see their faces clearly. While he was no lip reader, he was certain that whatever it was they were talking about was nothing good. He maintained his focus on them through the scope.

"Please, sit. Enjoy a rest in the sun," Krasnoff said, watching the men settle on the track bed.

"I know you've been wondering what this is all about. You know we're in Alaska, on American soil and I've told you that we will be conducting an unusually sensitive mission. So, I think you deserve more than the usual briefing. You should all know why we're here. The Americans' have had problems recently with what they call their 'nuclear surety' program. Several of their generals have been fired and the media has published the story. These nuclear handling mistakes have put additional pressure on the American president to reduce the numbers of their nuclear weapons. So, Moscow has decided to help this along by creating a situation to drive a great wave of urgency on the Americans to begin reducing their weapons now. Our mission has been to covertly penetrate U.S. territory, which we have done, and make this situation happen, which we are about to do," Krasnoff explained.

"Major Krasnoff, all of these men here have been engaged in dangerous missions throughout the world and on the territory of other nations. Exactly what are we to do here?" It was Captain Ivanov.

"Captain, a day and a half from now, a train will cross the border near here from Haines Junction on its way to Fairbanks. But several miles west of where we are now, it will leave this track and take a spur to Eielson Air Force Base, twenty-six miles southeast of Fairbanks. That U.S. air base owns several diesel locomotives which for years have been moving coal and other goods into the base, so the Americans have hoped that this train and several others like it would not be noticed, especially by imagery platforms." Krassnof looked up to the sky and the eyes of the others followed suit.

Krasnoff continued. "But we have noticed. This particular train will be moving ten ICBM nuclear missile warheads to the air base. Our mission is to stop the train, neutralize the engineer and the security force, puncture the diesel fuel tanks and crash the train off this bridge into the canyon below. Nuclear contamination of this wilderness park will follow."

"How will we stop the train, Major?" Lt. Sorokin asked.

"Before the train arrives, we will remove bolts, spikes and two sections of rails right at the edge of the bridge, where we are now sitting. I will be wearing an American Air Force cold-weather parka with a lieutenant colonel's rank on my cap. I will wave down the engineer, indicating trouble ahead on the track, ice expansion pushing against the rails or some similar lie. Lt. Sorokin, you, Mikulich and Sokolov have trained with the EMP rifles and are proficient with them, yes? So, before we leave the camp on mission day, you must make sure that the generator has fully charged your backpack and waist-pack batteries."

"Yes, sir." The three lieutenants nodded.

"So, I will approach the diesel's cab, explain the problem and the engineer leans out to talk to me, all three of you will open fire and strike him in the head with EMP bursts, stopping his heart. Lt. Bardinoff, you will be in the woods near me and will immediately climb up into the cab, and engage the brake to ensure locomotive doesn't move any further. If the security force tries to exit their car, engage them repeatedly with EMP bursts on their faces. You must hit skin for the shots to be effective. Once the security force is neutralized and contained, we will blow the seal on the door to the nuclear weapons container car, where Lt. Bardinoff will join me again to help remove an item and strap it to the sled. Any bodies other than the engineer will be placed back in the security car or the in weapons car. Lt. Bardinoff will return to the cab, set the engine's accelerator for maximum velocity and exit the locomotive. After the crash, we must leave the area quickly due to the rupture of the nuclear containers and immediate contamination."

"We then go back to the coast, sir?" Ivanov asked.

"You, Captain Ivanov, will take your team to the coast for immediate extraction. We will break camp and pack all supplies before beginning the mission. I will take one snow bike and the sled. I have another mission to perform... near the Yukon border."

"Five of us can't ride on two snow bikes, sir," Ivanov objected.

"I leave that to you to figure how you'll do it, Alexi. There is no other option. Men, we will all celebrate back in Moscow! The Americans won't even..." Krasnoff paused, the smile leaving his face as he stared across the bridge. The men scrambled to their feet, their assault weapons at their hips.

McClure swung his binoculars west. Two snowmobiles were approaching the Russians across the bridge, one blue and one orange, oblivious to the danger.

Up on the ridge, McClure bit back a cry. "Noo, no, no..." he moaned. "Robert, don't, go back. No." He watched as Krasnoff and Ivanov stepped forward in the front of their men as the snowmobiles pulled up.

Robert climbed off his snowmobile, grinning and with his hand extended. "Hello guys, hey, nice parkas, all coordinated too! Problem with the tracks?" he asked.

The only response was the metallic snap of four rifle slides chambering their rounds. Robert stopped in his tracks, the grin leaving his face abruptly.

"Hey, what's going on? Look, I didn't do anything! We'll go, no problem. Have a nice day." Robert turned away from Krasnoff.

"Stay right where you are or you will be shot dead where you stand. Understand?" Krasnoff's English betrayed no accent, but his harsh tone precluded any argument. Ivanov's head snapped toward him in surprise.

"Who are you and what do you want here?" Krasnoff demanded.

"I'm Robert Ewan. That's my son, John. And hey, nothing, we don't want anything. We're hunting, gonna swing over to the east, so we came across the bridge. We're just passing through."

"Call your son over here. Now!"

"No, please, please let him go. Keep me if you have to. John, you go ahead home, son, go now!" Robert shouted over his shoulder. Fear flitted across the boy's face.

Krasnoff snarled, "John, come over here next to your father. Now, John. If you try to escape on your snowmobile, you will be shot."

John climbed uncertainly off his snowmobile and walked toward them, stumbling a little on the tracks. "Please, mister, my father was just..."

"Shut your mouth, boy." Krasnoff turned back to Robert. "Do you have anything you want to tell me, anything at all? What kind of day you've had? What you've seen here?"

"Look, I won't tell anybody anything, I promise. Please just let us go. You can have the snowmobiles."

"Then, you have nothing to tell me, nothing to say?"

Robert's voice quavered and he thrust his hands out pleadingly. "Mister, please let us go. I... we didn't see anything here. I... um, maybe I do have something you'd want to know, if you promise to let us go."

"Really? Well, what is it?"

"You promise to let us go?"

"Mr. Ewan, we mean no harm to you or your son. You startled us, that's all. Of course you can go. So, what is it you have to tell me?"

"There's a federal officer, a special agent, not far from here. We ran into him a couple of days ago. He's armed." Robert's chest heaved, his heart pounding wildly. He had to tell them something, anything, to at least try save his son.

"A special agent? Where did you see him? Do you know where his camp is? Point the direction out for me."

McClure saw Robert's arm swing toward him, past him and beyond, to the south toward his camp. "Oh shit, Robert... no. Get out of there. Run. They're gonna kill you. Go, go." He shook his head, chewing on his bottom lip until it began to bleed.

Krasnoff turned to look and let loose a flood of Russian to the men. They all turned looked off in the direction Robert pointed. All except Captain Ivanov, who was still staring at the major. Even at this distance, peering through the scope, McClure could tell that Ivanov was upset by whatever the major had said.

Krasnoff swiveled on his feet, and give one last appraising glance at Robert and John, then stepped aside, nodding curtly to the waiting soldiers. "Now! Both!"

The earsplitting rattle of automatic fire shattered the silence and echoed through the canyon. In an instant, Robert and John were hurled backward, dozens of rounds ripping into their chests and heads and blood staining their parkas. Krasnoff raised his hand to halt the firing. Without further direction, the snow men picked up the limp bodies and tossed them off the bridge into the canyon. The snowmobiles followed, smashing apart on the rocky sides of the canyon.

From his vantage point, McClure watched in horror, shaking

with rage. He clenched the rifle in his fists to keep himself from blowing Krasnoff's head off. That would be a fatal mistake… he was outnumbered and far too close to risk a direct assault. His eyes closed in utter revulsion. A tear tracked its way down his cheek.

CHAPTER 21

Alaska

McClure's legs pumped furiously, hopping through the snow like a kangaroo on crack. His breathing was labored, his lungs shrieked for air, the muscles in his calves and thighs screamed in protest, but he kept pumping. He had to get back to camp, pack it up and get away from the snow men as fast as he could. He knew now these had to be the men who had murdered Jack Stroud. They had just slaughtered Robert Ewan and his son.

Stroud was paralyzed by fear and didn't have a clue what to do. Ewan was caught on a narrow bridge, with no warning and no options at all. No way would he let that happen to him. These Russian assholes weren't getting to Dave McClure. In two hours, he saw the opening in the trees that marked his campsite. The sun had slipped below the horizon, and darkness would soon overwhelm the forest.

He staggered into camp and collapsed in front of the tent. Still gasping for breath, he reached down and unsnapped the snowshoes and laid the rifle across his lap. His heart was pounding in his chest. *Slow down, Dave. Take it easy, think. Think, Dave.* Okay. He sat back on the spruce boughs, unzipped his parka to cool off, and gulped down what was left in the water bottle. Much better. McClure took a deep breath to collect himself.

The first thought was that he didn't blame Robert Ewan, not

even a little bit. The guy was trying to save his son. However, it appeared that Robert may have told the snow men about him and his camp. He had seen Ewan point in that direction. If that was true, then McClure had now lost his one singular advantage… surprise. Or had he? The Russians would also have to travel back down to their ravine, and in darkness. By the time they'd arrived, night would be upon them. After such a long trip, would they really want to try to find him tonight? In the middle of his musing, he suddenly realized that he'd skipped lunch, not intentionally but because he was so intent on watching what was happening on the bridge. He'd munch on some snacks for dinner, meager as they were.

A unexpected thrashing shifted his attention to his right, off in the woods. A frenetic, scurrying sound. McClure reached inside his parka and unsnapped the retainer, sliding the .44 magnum pistol from its holster. He eased up from the ground toward the noise. It wasn't night yet and he could still see fairly well. As he approached the stand of birch trees, he laughed out loud with relief.

A red fox was caught in one of the snares. He'd forgotten all about them. He pushed the revolver back down in the holster and snapped the retainer, still smiling. Blood peppered the snow around the fox, so it had injured itself in its frantic effort to get free. The fox turned and snapped at him as he drew close. McClure shook his head, and grinned again.

"What're you doing in my snare, you little shit? I see you ate the jerky though, didn't you? Take it easy now… I don't eat fox." McClure squatted about three feet away. The fox tried to lunge at him, but despite its bravado, he could see it trembling. It was scared to death. *Well, I wonder what we have in the other traps?* he thought. "You stay here and I'll be right back." The fox yelped.

He went to both of the other snares and saw that he had two rabbits. Both were good sized, and already dead. Rabbits were fragile creatures, these two had probably died out of sheer fright, struggling against the wire loops around their necks. They had just given up. He returned to the tent and dug out a black plastic trash bag from his pack. In short order, he dismantled the two snares with the rabbits, wound the wires around the stakes for storing, and dropped the rabbits in the bag.

He walked back to the fox, which was still feverishly trying to jump free of the snare's wire loop. "What am I going to do with you, huh?"

He squatted there watching the fox, holding the bag with the two rabbits in one hand and his rifle in the other. McClure's brain began to tick, and he sat back in the snow sorting through options. The fox was desperately fatigued, panting with exhaustion. It also eventually slumped to the snow, still eyeing McClure warily and snarling. He had two rabbits and a fox. How could he put them to good use, other than eating the rabbits and setting the fox free? Something was gnawing at the back of his mind. Two rabbits and a fox. And two humongous grizzlies were wandering through this area night and day. NIGHT… and day.

Dave abruptly sat up straight and his eyes widened at the thought. Trapping a grizzly was well beyond his ability and certainly not on his bucket list. So, what then? These three animals have a lot of blood and guts in them, don't they? He wet his index finger and stuck it up in the air. Wind from the south, onshore breezes gusting from the bay that would be running up through the Russians' camp in the ravine and north to the woods and the meadows beyond.

The idea crystallized in his mind. He wasn't going to wait for the snow men to come after him. No, he was going to hit them first, and hit them hard. They wouldn't remotely expect it tonight, and he would use the creatures of the forest to make it happen.

"Gees fella, I am really sorry about this. But it's something I have to do." The fox stood up, growled and again tried to leap at McClure as he rose to his feet. Two quick blows to the head with the stock of the rifle and the fox fell to the snow, its eyes lifeless and glazed over. "I'm truly sorry, little guy. I am. But, you are giving your life for your country." He nudged the fox with his boot but it didn't move. He knelt and removed the wires from the fox's neck, dropped the animal into the bag with the rabbits, and tied a knot in the top.

Back in camp, plunked down and checked his watch... 6:30 PM. The sun was down and the arctic night was up, a crystal clear night filled with stars. He reached into the tent for the sterno unit and set it up quickly. The snow melted and water was simmering for coffee in no time. He munched the jerky and crackers, then sat sipping the coffee and watching the forest.

Methodically, he examined the pistol and rifle barrels, satisfied that they were clean and clear of debris. Using the sanding block, he honed his SOG knife blade to a razor edge. He had time. He'd leave here about one o'clock in the morning, be there a little after two o'clock, circle around to enter the camp from the south. His Russian wasn't fantastic, but it was passable. He could feel a mix of energy, excitement and fear coursing through him.

McClure looked up, fixing his eyes on the stars. "Anne, sweetheart, if you're listening, you know this is something that I have to do," he whispered, his face pale from the faint light of the sterno. "I love you and Michael, and I always will. I will always treasure our love. But, I think you understand, I have to do this. It's who I am, sweetheart. Good night, baby." His chest heaved as he let out a heavy sigh.

He let the sterno stove burn on. It gave a little light and he found the flutter of the flames comforting. The Union Pacific Railroad slipped into his consciousness. The Union Pacific operated up here, he knew that, but how did he know it? What was so important about the track from Haines Junction to Fairbanks? Something was buried in his memory. His eyes slowly lit up... Peacekeeper Rail Garrison. That was it.

In times of an international crisis, ICBM's on train cars would be dispersed from a garrison base out into the wilderness where they could launch from a pre-designated point against an aggressor. The Russians couldn't keep track of the trains, could never count on their ability to geographically localize them, just like the 'boomers', the nuclear subs. That randomness served to deter a preemptive first strike by further complicating the Russians' nuclear target planning. But that initiative was scrapped after the Iron Curtain came down in November 1989. So, what else?

McClure made another cup of coffee, his last of the instant mix. The thought was driving him nuts, right on the edge of his consciousness. Why was this Russian special forces unit here? Obviously, their plan involved the track and the bridge. Ever so slowly, the memory emerged. Maybe five years ago. He wasn't directly involved, but a couple of agents in the office were detailed up here and bitched about the assignment. He could feel himself getting close. Yes, the agents were detailed to conduct a Counterintelligence Risk Assessment of the tracks through Alberta, British Columbia, the Yukon, Haines Junction and then to... no, not Fairbanks. Eielson Air Force Base southeast of Fairbanks.

McClure remembered the agents had complained for days after their return about the harsh working and living environments out here. What had been going on? The U.S. Air Force and Eielson Air Force Base were doing something classified way beyond Top Secret. The program was heavily 'compartmented', and then even more restricted. Access wasn't just for personnel with a "need to know" requirement, it was an absolute "must know."

Trains would come across this track like normal rail traffic to Fairbanks, but then were switched off on a spur to Eielson. What was important about that? Something about... North Korea? Yes! The sudden realization brought Dave to his feet, breathing fast, his fists clenched and punching the air.

"Shit! No, dammit! You assholes aren't getting away with this! No fucking way!" he thundered.

This decision was no longer a 'want-to-do', but a 'must-do'. His face set, McClure packed away the sterno and pulled his backpack out of the tent. He rolled up the sleeping bag and stuffed it in the pack, lots of space now with the food gone. Next he disassembled the tent, rolled it up tightly and strapped it to the bottom of his pack. After his visit to the Russians, he'd come back here, grab his gear and move camp somewhere to the northwest. And he'd have to do all of it in darkness.

A smile broadened his lips, the first one all day. He looked at his watch... 8:00 PM. He had some time to think, to prep, but he decided to use that time to move closer, very carefully and quietly. He couldn't sit around for hours doing nothing, he'd go crazy with anxiety. He shoved his pack over his head and up into the air, looping it on a branch as high as he could reach, then gathered his gear and stepped from the campsite into the black woods.

What was it he'd once read that Sun Tzu had said? *"Take the enemy unaware by surprise attacks where he is unprepared. Hit him suddenly...."*

Yes. McClure was going to give these sonsofbitches a war they would never forget. It was time for these snow men to account for Jack Stroud, Robert Ewan and John. If luck was with him, they'd never leave these shores alive. And, if it turned out he didn't survive the fight that was coming, at least he'd cripple these bastards so bad that they'd have to abort their mission. The thought of a court trial never once entered his mind.

CHAPTER 22

Northern Virginia

Even flying in first class, the air time from Tel Aviv to Washington, DC was brutal. Taken back to back, the El Al 747 flight to Zurich and connecting Swissair 767 flight to Dulles International were a lot rougher than they expected. Raphael and Meira were drained. It was right about 8:30 PM when they grabbed their coats and carry-ons from the overhead bin, and staggered up the ramp into the gate area. Two conspicuous guys in suits stood stiff up front.

"Mr. Mahler, Ms. Dantzig?" the man in the dark navy top-coat queried.

"That obvious, huh?" Raphael answered, forcing a smile.

"No, you both just look professional, that's all. You should see the riff-raff that comes into Dulles from Europe, Africa, and Asia, all looking like bums. I find it hard to believe that people travel internationally dressed like that. Anyway, it's a just short walk over to the other hangar. I'm Walter, by the way."

"I get what you're saying. I never thought about it. Ahhh, Walter your true name?"

The two men looked at each other and laughed. "Yes, Mr. Mahler, it is. I'm Walter Mackenzie and this is Bill Farris. We're not undercover here, there's no op going on. Let's walk and talk, shall we? Need any help with the totes?" Walter continued to chuckle as they turned left and walked through the concourse.

"No thanks, we're fine. I'm Raphael and this is Meira." Meira nodded, too tired to speak. "What about our luggage?"

Walter smiled as he strode briskly along. "Raphael, Meira, I'm pleased to meet you. We have people getting your bags off the plane now. They should actually be at the hangar before we are. By the way, Mr. Novak came out here himself tonight. He said to tell you that he's very eager to meet you."

"Thank you, Walter. That's quite an honor. We were in on the phone conversation between Mr. Perutz and Mr. Novak yesterday. He sounds like a great guy."

"He is. Everybody at Langley Center likes him. Novak's a super guy. He looks out for his people. That's more important to us than salary."

"I can understand that. Walter, Meira and I are quite tired. I hope there's no heavy itinerary this evening." Raphael stifled a yawn.

"No, none at all. Mr. Novak just wanted to be here to meet you and welcome you to Washington and the United States. We're taking you straight to the hotel where you can get some rest. First time here?"

"Yes."

"Well, you're going to like the Ritz Carlton. It's a classy place, good food in the bar bistro, it's called ENTYSE's or something like that. Right off the lobby. Oh, and the hotel is right behind the Tysons Galleria mall, and only a short ride over to the Company. We checked... you're already registered. You have a couple of nice junior suites."

They were approaching a formidable looking security door flanked by two armed guards in black uniforms. One of them smiled, nodding to Walter and shoving the door open. Through the door, down a short corridor, and another door with two more guards. They ambled through the last door and finally entered a cavernous hangar with three jets parked, all bearing the American flag on their tails. Meira looked at Walter.

"The guards, were they…? she asked.

"U.S. Secret Service, Uniformed Division. Good guys. They supplement the TSA, Customs and Homeland Security people out here. But really, they're only here for this hangar. The diplomatic fleet is out at Andrews Air Force Base in Suitland, Maryland, just across the Woodrow Wilson Bridge, but some special air flights go out of here. So, we have the Secret Service here as well."

"Great, thanks. Just curious." Meira said.

They looked ahead and saw another group of three men and a woman approaching. A distinguished looking man in a tan cashmere topcoat walked quickly toward them. He had a smile on his face and sandy hair with a little gray on the sides.

"Hi Raphael, Meira… I'm Pete Novak." The rest of the entourage turned and headed straight for three black Cadillac limos parked just inside the hangar.

"Mr. Novak, it's so very good to meet you!" Raphael enthused, shaking Novak's hand vigorously.

Together they walked toward the limos, Novak leading the way. "You folks must be beat. We're taking you straight to the hotel in Tysons Corner, right next to our front door in McLean. It's a long way from Tel Aviv to Washington, isn't it?"

"Yes, sir, in more ways than one." Raphael answered.

Novak stopped in his tracks and turned to Raphael who blanched instantly, thinking his remark was inappropriate, uncalled for. Meira's eyebrows went up, wondering what was coming next.

Novak's face and voice were resolute. "Raphael, you have wisdom beyond your years. I'm going to enjoy talking to you and working with you. I want you both to know one thing up front. You can, as we Americans say, take this to the bank. No matter how screwed up our nations' politics, our foreign policies and our relations seem to be, we ourselves, the Agency and Mossad, are brothers and sisters in the same war. Despite the problems our respective leaders may have with one another, I and my personnel will always be there to support Mossad and the nation of Israel. And we expect that you will be there for us. Never forget that."

"Sir, that's very good to hear. I meant no criticism with that remark." Raphael's face was somewhat pale.

Meira stood in place mesmerized by Novak's very presence and his calm, confident tone. *This is a good man,* she thought.

Novak smiled. "Good. Please know that I'm very glad you're here. From what Taavi said, we have some very serious matters to attend to. Walter and Bill will take you to the hotel now. Have something to eat, maybe a little something from the bar to help you sleep. They'll pick you up in the morning and bring you over to my office. You'll have non-escort badges, so you'll also have some freedom while you're here. I'm planning at least one conference call in the afternoon, after we talk and I'd like you to be in my office for that. That's pretty much it for now. Questions?"

Raphael shook his head. "Not at this point, Mr. Novak. Thanks for everything." Meira shook her head as well.

"Great. Ah, weapons. You have your service weapons with you, ammo? Just for the remote possibility that we have to go operational."

"Yes, sir, we do." Raphael answered.

"Okay then, see you both in the morning." Novak shook hands with both of them and climbed into the rear seat of the nearest limo which abruptly pulled away toward the open hangar doors.

Walter stepped into the driver's seat of the next limo as Bill slipped into the front passenger seat. Raphael and Meira climbed wearily in back. The limo jolted forward, and then stopped as Meira spoke up.

"Our bags, Walter?"

"In the trunk, Meira."

"Gees, you guys are good," she chuckled, leaving all of them smiling.

Heading east, the limo swung out on the Dulles Access Road and after a quick two miles, turned onto the Dulles Toll Road. As tired as they were, Raphael's and Meira's faces were pasted against their windows, watching all the lights and sights. Reston, Virginia was gorgeous, lit up at night. The drive on the toll road took about ten minutes, after which the limo turned onto Spring Hill Road, which morphed into International Drive. Minutes later, they made a left onto Tysons Boulevard and pulled in front of the Ritz Carlton. All four exited at once, and Bill immediately walked to the rear to pull the luggage from the trunk and pushed them over to a bell hop.

"You folks get a good night's sleep. We'll see you right out here again at 07:30 AM tomorrow. You'll find a couple of great restaurants in the Galleria… Maggiano's, Legal Seafood, the Capital Grill, to name a few. I think you can enter the mall right through the back of the hotel. Ask the desk about that. Have a good evening," Walter smiled.

"Walter, Bill, thanks so much for everything. See you in the morning." Raphael answered. They all shook hands, then Raphael and Meira watched the shiny black limo motor out of sight.

"Wow," Meira gasped.

"Wow is right. First class. We have to remember to tell Ari and Taavi about the welcome they gave us. And Novak looks to be a super guy. Let's go in, get our keys, throw our stuff in the rooms and meet down here at the bistro they mentioned for a drink and something to eat. I don't feel like walking anywhere. You?"

"Agreed. I'm pooped."

It took them about twenty minutes to check in, drop their luggage on the beds, wash up and head back downstairs. The junior suites were side by side, spacious and attractively appointed. The windows featured stunning views of Northern Virginia. They met up outside the lobby's ENTYSE Wine Bar & Lounge, an impressively nice place to enjoy a drink and superb cuisine.

Raphael asked for a vodka gimlet on the rocks with Stolichnaya vodka. Meira sipped a glass of chardonnay. They both ordered salads and the soup du jour, a wonderful crab and lobster bisque with a dash of sherry. They managed some small talk, but the emphasis was on the meal. They were starved and the food was absolutely delectable.

"My, that was scrumptious, wasn't it?" Meira breathed.

"Oh yeah!" Raphael sipped his gimlet.

"That looks good, Rafi. Can I have a taste?"

"Of course." He pushed the drink toward her.

"Wow. It's vodka and what else?"

"Vodka and Rose's Lime Juice. It's a sweetened, syrupy lime juice. Great with vodka."

"I really like that. Now I have a new drink to order, thank you very much. What about dessert? We're on per diem."

"Ah, okay, but we share. How about two coffees and a crème brulée?" Raphael asked.

"Okay, we share. Coffee and crème brulée sounds yummy."

They sipped their coffees and leisurely dipped their spoons into the crème brulée, savoring the taste. Raphael looked at his watch… 10:30 PM. Meira noticed.

"In a hurry?"

"No, but I'm whooped and we have a lot to do tomorrow."

"Rafi, you and I are in America. Tomorrow we will be at CIA, Langley Center. This is beyond my wildest dreams. Isn't it wonderful?"

"It is fantastic, Meira. But we have to remember why we're here. Some very serious business."

"Does that mean you don't want to play house?"

"You have the most unique way of putting things. We can't, not tonight. It'll be midnight soon, and we've got to shower and get some sleep."

"Shower? Oh, could we just…?"

"Not tonight."

"You're a poop."

"Right. I'm beginning to think that's my nickname."

"Okay, but you can't run away forever. Remember, no walls."

"Come on, I'll walk you to your door. To - your - door." Rafi laughed, and so did Meira.

CHAPTER 23

Alaska

A night black as coal consumed the forest. The moon scraped along the horizon and finally dipped below it. McClure moved through the woods as slowly and cautiously as was possible. He knew it wasn't far to the open snow field where there'd be a lot more light, but he had to admit it was a little spooky moving through the forest at this hour. His skin tingled, and he could feel a cold sweat moving up his neck. *Come on, Dave, settle down. Think about your days in the field operations school. Concentrate. No time back then to be an old worry wart, and no time for that now. Chill out.*

In ten short minutes, he advanced through the tree line and took a deep breath. The moon was gone, but lots of stars. The reflecting light from the snow field was soft and watery, but it was enough. Even that small amount was a relief. McClure turned south and found a slow rhythm with the trekking sticks, back and forth, carefully avoiding any audible clicks, and planting the snowshoes as quietly as possible. His breathing slowed to match his easy pace. The forest, the snow field, the wilderness itself, was soundless, and in that way, surreal. He took his time, keeping a light, smooth stride and pivoting his head frequently left and right, his senses on high alert.

His eyes widened when he saw them coming from the

snowbound meadow in the east and west toward the forest. He'd thought of the possibility, in fact he'd actually been hoping for it. On a clear, still night like this, why wouldn't they be out? Either Bruno or Brutus. The tracks were close together, so the grizzly was going at an easy pace and taking his time.

McClure bent down to take a closer look. The prints were enormous! Fresh as hell, crisp around the edges and a little crumbly… an hour ago, maybe two. He felt the gooseflesh rise on his arms, remembering all too well the confrontation with the huge bear that had given him cause to clean his underwear. It was indelibly etched on his brain.

His eyes darted around unceasingly, but saw no more bear sign. He checked his watch… almost 10:30 PM. He felt he was close, but the tediously slow tempo made it impossible to be certain. A few minutes later, he paused again, recognizing the entrance through the black spruce to his overlook position. He decided he'd go take a gander.

With utmost attentiveness and care, McClure eased himself down in the soft snow and removed his snowshoes. Squirming on his belly, he stopped frequently to listen on his way to the ravine's edge. He gently pulled out his binoculars out of parka and swept his focus across the encampment. Faint curls of smoke rose from the fire pit. An occasional cough came from the tents.

The snow men had returned from their sojourn up north after murdering the Ewans. They had slipped into an uneasy sleep, gathering as much rest as they could for what they were anticipating tomorrow. But the man with the binoculars was betting a change of plans was in store for them.

He pushed himself backward into the little clearing and sat upright. Without really thinking about it, he pulled the compass out of a pocket and, with no exact purpose in mind, took his bearings. He glanced up at the night sky, a black cushion clustered with millions of bright, silver pins. Hell, he couldn't identify any constellations except the Big Dipper and the Little Dipper. And, this close to the Arctic Circle, even that information was of no use he could fathom.

Between the trees, he glimpsed the expanse of the snow field, a sea of shimmering light that swelled on to the horizon. It was a little after 11:00 PM. He had two hours before he would need to move south. He wouldn't go sooner, he didn't want to put his scent on the onshore breezes.

McClure's mind wandered through a maze of memories. In the dark and noiseless woods, every thought was so vivid it seemed to shout at him. He remembered sitting in class at the covert site for part one of the CIA field operations school, listening to his favorite instructor, J.T. Brannon. After that experience, his FBI training down at the Federal Law Enforcement Training Center at Quantico seemed fairly benign. In many ways, it was. The FBI wasn't about teaching you how to approach in total silence a guard walking a perimeter patrol, place a wire garrote around his neck, crush their windpipe and then finish them off with a knife thrust to the carotid artery.

His mind moved on. He remembered meeting Anne in the shrimp and oyster bar at the Fish Market restaurant in Old Town Alexandria. After a drink or two and some small talk, they shared a platter. He remembered the soft glimmer of her eyes looking into his. He remembered how later, after dinner, they walked hand in hand down King Street and along the placid water of the Potomac, and back up the hill where they danced to some Irish Celtic music at Murphy's.

By the stroke of midnight, in just five brief hours, they were thinking that they could dance together forever. He remembered sending a young Michael full of energy way too deep to catch his pass and Michael stretching for the ball and falling on the grass laughing. He remembered.

McClure paused in mid-memory. *How about some future*

think, Dave? What was he going to do when he got out of here? He was an incorrigible optimist, always had been, and felt in his heart that he would indeed somehow survive this ordeal. And then what? Go back to the FBI? Take Pete Novak up on his offer to return to the Company? Well, before he seriously considered any of that, he needed to attend to the matter at hand. He peeked at his watch… 1:00 AM. Did he really have that many memories? Yes, those and a million more.

The air was frigid, probably down to minus 10 degrees, but he hadn't noticed the cold. The sky was still black, with millions of twinkling dots stretched across it. He could feel a soft breeze gliding up from the coast. No heavy wind. He began to stretch and move around, clearly hearing the silent whistle of the referee… game time.

Moving soundlessly down to the southern entrance of the ravine might take as much as 45 minutes. So, with the same care as before, he strapped on his snowshoes and stood up, slipping the rifle over his shoulder and retrieving the trekking sticks. McClure began ever so quietly tiptoeing out through the trees and into the open field. He could see the bay. Jagged white icebergs floated listlessly on the dark water.

He trudged along and before long reached the shoreline. To his right, he saw the covered rafts, ghostly pale in the dim light. He knelt down and stared out over the bay, sweeping his binoculars across the surface… nothing there. Lurking under the clumps of ice was a Russian sub, the Spetsnaz team's ticket home. McClure would see about that.

His watch read 02:00 AM. Time to execute plan A, though he really had no plan B. The trail from the shore north to the encampment appeared well trodden, so the snowshoes would be more of a hindrance than a help. He sat on a covered raft to remove them, using the carry strap to sling them across his back, then collapsed the trekking sticks and hooked them to his belt. He was glad the moon had set, leaving only an ashen reflection on the snow.

With the rifle in his hands and at the ready, he picked his way up into the ravine. He tried to walk in the compacted footprints to avoid crunching down on new snow. In a half hour, he began to see the outlines of the tents. The larger equipment tent appeared first in line, its flap conveniently open. McClure moved silently up to the entrance and saw the transmitter in its zip cover on the ground just inside.

Stepping cautiously in, he gripped the cover's loop handles and lifted gently to judge the heft. It was heavy and would take a good deal of effort, but he was confident he could tote it back down to the shore. He glanced at the snow bikes and noted other open boxes lined up along the tent walls. Examining some of them, he was pleased to see food packs of stews, soups, even coffee. And oh joy, a box of Hershey bars! He would definitely return here before leaving the camp.

He stepped back outside and crept up to the first occupied tent, also on the right. Placing the rifle gently on the ground, he quietly drew his knife and slit open the back plastic bag that held the carcasses. It seemed logical if a grizzly were to enter the camp, it would do so from the north end of the ravine rather than the south. Spreading the blood, entrails and carcasses here would draw the grizzly right through the entire camp. The ensuing havoc would most likely be horrifically violent.

McClure took off his gloves and stuffed them in a pocket. He pulled out a rabbit and held it over the slope of the tent's roof, slipped his razor-edged knife into its belly, and slit it open from neck to tail. Congealed blood slowly dripped out onto the tent. He scooped out the entrails and laid them gently on the edge of the roof, resting the now very light rabbit carcass on top. He followed suit with the second rabbit. The fox was much heavier and more difficult to handle, but yielded far more blood and entrails. The fox carcass went on the ground next to the tent.

He stuck his hands in the snow behind the tent to clean them of blood, wincing at the burning sensation from the icy snow. The air was so frigid it was painful to take in a deep breath. He could imagine seeing his hands turn blue. He gently folded the plastic bag and stuffed it in a pocket. Gratefully donning his gloves, he abruptly paused, holding his breath.

"*Kto zdes?*" "Who's there?" Ivan Sokolov spoke from just inside the tent wall that was dripping with blood and guts.

McClure froze, dreading the idea of having to speak in Russian, but he had no option. "*Mne nado V Tu Alet! Tv Idi Spat!*" "I'm taking a piss! Go back to sleep!" he growled back. His face tightened, waiting a reply.

"*Ladno, biyastra!*" "Okay, hurry up!" Sokolov answered, obviously not happy that he'd been awakened.

"*Ladno.*" "Okay."

McClure let out a sigh of relief, slowly picked up the rifle, and stepped lightly away from the tent. He backed away cautiously, eyeing the entrances to the occupied tents, but there was no other sound or movement. Returning to the large tent, he again peered into the open boxes and smiled. He stuffed his pockets with enough food packs and Hershey bars to get him through another week. Moving over to the transmitter, he realized he would need a free shoulder to carry it, so he slipped the rifle quietly over the snowshoes on his back. He lifted one of the transmitter handles over his strong right shoulder, grimacing at the weight of it. Thank God he didn't have far to go.

McClure walked cautiously away from the camp. The trail to the shoreline was much easier going down the incline. Nevertheless, the weight of the transmitter dug into his shoulder and ached like hell. He had to fight against wheezing until he was out of ear shot. Out of sight of the camp, he set the transmitter down and moved the rifle over to his right shoulder. Grabbing both handles, he lifted the transmitter as far up his chest as possible and continued trudging down the trail.

At the water's edge, McClure set the transmitter down with a grunt and sat on a raft to catch his breath, his rifle across his thighs. He checked his watch... 03:20 AM. He had to get moving, but maybe there was time for a five minute break. All this wilderness trekking had put him in the best shape he'd been in for years. He was certain even the troublesome extra five pounds around his waist had disappeared as well. Still, carrying the heavy transmitter down with the rifle and snowshoes had taken his wind away. Gulping down some water rehydrated and helped him slow his breathing.

Reasonably rested, Dave stood and yanked the raft covers off, then did the same to the other raft. Pulling his knife from its sheath, he hefted the blade before plunging it into the thick rubber wall and slashing the length of each raft. He was surprised at how heavy they were, even deflated. Grunting with the weight, he dragged each raft to the ice bound rocks at the shore's edge and heaved them in. They sank slowly into the dark water, although trapped air bubbles kept them from disappearing altogether. Good enough.

McClure strapped on his snowshoes, reached over and hoisted the transmitter again, hauling it northwest along the icy shore. Where the snow was hard and crusty, it was easier to just drag it behind him. After maybe a half hour, he spotted a little spit of rock jutting about thirty yards out into the bay. He unstrapped the snowshoes, set the rifle on them, and tugged the transmitter out to the farthest point. The cover he tossed out into the bay. Then with a monumental effort, he flung the transmitter out into the water. It vanished with a satisfying splash.

His lungs gasping for air, Dave stumbled back along the shore and sat down by the snow shoes. His watch read 04:15 AM. So far, so good. The snow men now had no way to communicate with their sub, and no way to get to the sub out in the bay. After swigging more water, he began traipsing his way northwest.

An hour into it, he figured it was safe to start a generally easterly arc, the last curve in a large loop that would swing him back to his old campsite. Dave checked his compass and took his bearings, amazed at his new level of confidence for navigating through the wilderness. The butterflies in his stomach were gone, replaced with a feeling of lightness. At the same time, he remembered the old adage that confidence was all too often the opiate of fools.

It was nearly 06:00 AM when he spotted the large lump of his backpack in the trees. Without that marker, he might have trekked right by it. He reached up and shimmied it loose from the tree, letting it plop down in the snow. Too bad the Russians didn't have any jerky. With nothing in his stomach, that would taste awfully good right about now. His energy was once again drained. He felt as though he was running on batteries.

He stuffed more snow in his water bottle, slung on the backpack and shouldered the rifle. Using his sticks at a slow but steady pace, he plodded out, heading in the same direction from which he'd arrived. He had several miles to go, mostly to the northwest. And when he finally felt safe, his first priority would be to set the tent up and crawl in to catch some much needed sleep.

Four and a half miles north of the Russian encampment, foraging in a stand of birch trees, Brutus hesitated and lifted his massive head, sniffing at the wind. He heaved his tremendous bulk up on his hind legs and from the height of nine feet, sniffed again. No mistaking the scent. Fresh blood. The grizzly dropped to all fours, spun about and lumbered southward, where he felt certain he would find his next meal.

CHAPTER 24

McLean, Virginia

Meira and Raphael walked out of the Ritz Carlton at 07:30 AM to see Walter Mackenzie and Bill Farris already waiting, leaning against the sleek black Cadillac, talking and sipping coffee. Raphael waved and Mackenzie stepped forward, waving back. Raphael and Meira hurried down the stairs and over to the parked limo. They shook hands all around. Farris opened the rear door for them, and they slid in with a nod of thanks. As they fastened their seatbelts, the limo swung away and left onto Tysons Boulevard. Farris turned and put an elbow over his seatback.

"Did you already have something to eat?" he asked.

"Yes, thanks. We both had coffee and croissants," Raphael replied.

"You were right," Meira said, "That ENTYSE bistro was great. We had dinner there last night too."

"Good. Glad you liked it." Farris smiled.

"Well, just sit back and enjoy the scenery folks," Mackenzie said over his shoulder. We'll only be going about five miles. I could take an even shorter route, in mileage that is, but this time of day the traffic congestion around Tysons Corner is miserable. We'd be stuck in it for a half hour. This way is much faster."

Meira and Raphael were astonished. November, and everything was still so green. After only a half mile or so, the Cadillac turned right onto International Drive. After another right, the limo continued two miles down Lewinville Road and took a short jaunt on Chain Bridge Road where impressive, manicured lawns fronting brick colonial homes lined both sides of the streets. Finally, the limo turned left onto Dolly Madison Boulevard and left again into the Agency's visitor entrance. Walter maneuvered the Cadillac slightly right into the Visitor Control area, where a stony-faced security guard nodded at the escorts' IDs and checked Raphael and Meira's names off the list. Walter then steered over to visitor/VIP parking.

"It's November. How come everything is still so green?" Meira asked as Farris opened her door.

"It's been that kind of year, really. A long, hot and very humid summer with lots and lots of rain. We didn't think August was ever going to end. The cold just hasn't kicked in yet here," Mackenzie replied, sliding out the driver's side.

Bill Farris added, "You folks should really come back here when autumn is in its full splendor in Northern Virginia. You wouldn't believe how gorgeous it is."

"Well, if we only could. To do this again though, we'd need an invitation from your side," Meira laughed.

They stepped into the spacious lobby, their footsteps echoing off the marble floors, walls and columns. Raphael and Meira could scarcely take it all in. Speechless, she pointed to one wall with engraving in the marble, a U.S. flag was anchored on one side and a royal blue flag bearing the CIA seal on the other.

"That's the Honor Wall." Walter offered. "It's a memorial dedicated to CIA personnel who gave their lives in the service of their country. We pass it every day."

"Gees, that's impressive." Meira said.

"What's that over there?" Raphael asked, turning and pointing directly ahead.

Walter turned to look. "Oh, that, go take a look. It's a quote from the Bible, John 8:32… 'And ye shall know the truth, and the truth shall set you free.' That quote is really about who we are as an intelligence agency and what we do. Personally, I think it sums it up well."

Meira and Raphael walked over and lightly touched the engraving. Both were smiling as they turned back to Walter.

Meira nodded thoughtfully. "Wow. That's very powerful."

"I agree. It is. Let's get you both checked in and badged. How about a fresh cup of coffee before we head up to Novak's office?"

"Thanks, that sounds great, Bill. Meira?"

"Me too."

"Walter, I'll give Cathy a call and see if Novak is ready for us. You go ahead and take them in the cafeteria."

"Sure. Thanks, Bill."

The cafeteria was almost as impressive as the lobby with large windows overlooking a well-kept expanse of trees and lawn. Walter walked them through the line and paid for the coffees, then led them to a table by the windows so they could take in the view.

"Your facility is super, Walter," Meira said, gazing admiringly through the glass.

"Thanks. So's yours, especially considering your climate."

"You've been to Mossad Center in Tel Aviv, Walter?" Raphael asked, surprised.

"Yes, once, three years ago. We were checking up on a source who reportedly had some connections to Hezbollah. Examining prospects for doubling him over, but it didn't pan out as we expected. However, we did make the trip out there to talk to some of your folks. I loved your location looking out to the Mediterranean. Pretty neat. It was July though. Hell, and I thought the heat in Washington was bad in summer. Whew!" Walter laughed, turning his head as Farris approached.

"Novak said to come on up," Bill announced. They rose and headed for the elevator.

Minutes later, Cathy greeted them as they walked into Pete Novak's outer office. "Mr. Novak said to go right in when you arrived." She buzzed the door, throwing the electronic bolt.

Novak rose from his desk as they entered. "Hey guys, how're you doing? Please, have a seat. Ah, Walter, Bill, thanks very much for all the help. Please stay in the building where I can reach you." His hair was slightly disheveled for this early in the day, although the dark navy pinstripe suit and a flashy gold tie made up for it.

"Will do," Walter answered as he and Farris left Novak's office, quietly closing the door behind them.

Novak sat down, his hands folded on his desk, his eyes shifting from Meira to Raphael. "Well, to the issue at hand. Taavi and you both related your suspicions regarding a potential theft of a Minuteman III nuclear warhead. I know he wouldn't have alerted us unless it was serious, at least the implications seem to be. Taavi also told me that you both had led surveillance teams on Bizhan Madani and Leonide Krasnoff in Paris. The Russian mafia and Vevak, now I would agree that's an interesting, even disturbing combination. So, what you gleaned from that surveillance started all this. There was the transfer of the case at the Café Procope and the subsequent deposit of $25 million in a Zurich account. We looked into that and it checks out. The rest of it, though, involved lip-reading by one of your agents in the restaurant. He picked up on the word 'MIRV' and the number '450', and yes, I would say that's of concern. But we all know that lip-reading inherently has a large margin for error. Yes? What say you?"

"I don't think our surveillance erred either in what we picked up or what we pieced together afterward. Just too many coincidences," Raphael answered.

"We don't believe in coincidences either. In our business, we can't afford to. But I have to ask, is there anything else?" Novak pressed.

"The meeting in Amsterdam."

"What meeting in Amsterdam? When? Who?" Novak looked at each of them in turn, puzzled.

"Taavi hasn't told you this? Well, I guess he knew we were coming here to meet with you, that's probably why. Anyway, after the Paris meeting, Krasnoff had a meeting in Amsterdam. Viktor Nechayev flew in from Moscow to see him. The meeting location was obviously chosen very carefully and security was high, so we couldn't get a team in close enough to cover it."

"Viktor Nechayev? Nechayev came out of Russia himself to meet with Krasnoff in Amsterdam? Shit. Pardon my French." Novak paled visibly, his brow furrowed.

"Yes, sir. Mr. Perutz was greatly concerned about that and felt it was time to call you."

"Yeah, no kidding. I'm glad now you did. Thanks." Novak paused, a thoughtful look washing across his face. He reached over and pressed his intercom button. "Cathy?"

"Yes, sir?"

"Ask Joe Moretti if he can stop what he's doing and come up here. I've got a couple questions for him."

"Yes, sir."

Novak looked back at Raphael and Meira. "Joe Moretti heads up our strategic deterrence team. I hired him away from the Missile Defense Agency over at Fort Belvoir. MDA evolved from the old BMDO... ah, the Ballistic Missile Defense Organization. Joe's very knowledgeable."

"Mr. Novak, before we came out, we questioned one of our missile technology experts about this." Meira interjected.

"That's good, Meira, but I'd like to ask Joe a couple of questions as well. Maybe we'll get some validation."

"Yes, sir. Actually, I think we would very much like to hear what he has to say," Raphael broke in, eyeing Meira.

The office door buzzed and a craggy, fifty-ish gentleman with a substantial paunch and wearing cross-trainer shoes walked in. He nodded to Novak, Raphael and Meira in turn.

"Mr. Novak, you wanted to see me?"

"Yes, Joe, please take a seat. I have a couple of questions for you. You have a few minutes?"

"Yes, sir. Fire away."

"The topic is Iran, Joe. What do the Iranians have for missiles, anything strategic?"

"Iran, well, let's see. We know they've been making some upgrades to their Safir missile; that's their largest. It's a very capable, qualified launch system. They could launch a Safir missile successfully against most any target in SW Asia, the Middle East or North Africa. That's about the range limit of it though."

"Hmm... Safir missile. Joe, another question, a blunt one. Could they... would it be possible to put a Minuteman III MIRV nuclear warhead on a Safir? That's about the only way I can put it, Joe."

"A Minuteman III MIRV nuclear warhead? Well, if I recall, the Safir is somewhat larger than the Minuteman III. I suppose it could comfortably accept a payload like that; it's well within the parameters. Sir, I'd have to say yes, it's technically feasible. They most certainly have the technology. That is, of course, if they could get hold of one, but that's another question." Moretti looked at Novak with concern.

"Indeed it is, Joe. That's another question. Well, that's all I really wanted to ask you. Thanks for coming up. Say hi to Cheryl for me. Oh, and please keep this discussion to yourself, all right?"

"Yes sir, I will. Have a good day." Moretti stood and walked to the door, throwing a curious look at Raphael and Meira as he left.

Novak leaned forward, resting his chin on his clasped hands. "Gees, there it is. Taavi said that Leonide Krasnoff has a brother named Vassily in the GRU, and that he's detailed to some Spetsnaz unit. All right, Raphael, Meira, I'm with you. I'm very concerned."

Novak paused, then abruptly pushed the intercom button. "Cathy?"

"Yes, sir?"

"The secure teleconference call we had set up for this afternoon… I'd like to move it up. Fifteen minutes from now if possible. Go Flash Override on this if you have to."

"Yes, sir."

"So, get General Forrester, that's Wallace Forrester, at Global Strike Command down at Barksdale Air Force Base near Shreveport, Louisiana. Also, General Kenneth Cantwell and Brigadier General Dick Kean. Cantwell's out at Strategic Command, STRATCOM, at Offutt Air Force Base in Omaha. Kean's the AFOSI Director; he's down at Quantico. I want Kean because he may want to step up their counterintelligence support to a couple of our bases. Oh, and see if you can get Lt. General Charles Durwell, Strategic Deterrence over on the Air Staff at the Pentagon. Did you get all that?"

"Yes, sir. I've got it."

"Please see if you can get them all in fifteen minutes, Cathy. Oh, and another thing, call Walter Mackenzie and see if he can run down the director for me. I may have to brief him in an hour or so."

"Will do, sir. This all sounds pretty serious, Mr. Novak. You know you're giving me goose bumps."

"Well, you're in good company, Cathy. I've already got them."

Novak gave Raphael and Meira a tight smile. "Let's go

freshen up our coffee, shall we?"

He stood and the three of them walked out to the coffee machine the outer office. Their cups refilled, Raphael and Meira ambled around the office, looking at photos and other memorabilia that covered the walls.

"That's my 'I love me' stuff… forty years of hobnobbing with counterparts all around the globe," Novak smiled.

"Mr. Novak, you've… shall I say… been around," Meira said, impressed.

"Well, it certainly feels like it." He pointed to a photo in the corner. "That's me as a youngster in operations chewing on some beef barbecue out at the Reagan ranch in California. Looking back now, I have to say that I can't believe I was actually there."

"I understand, sir. That's sort of how we feel now being here at CIA," Meira admitted.

"Yeah, I can see that." His smile faded. "Let's go back and get ready to hook into this call." He swiveled his eyes to Cathy and she gave him the thumbs-up signal.

Back in his office, the three seated themselves around Novak's desk. He pressed the speaker. "We wired, Cathy?"

"Yes, sir, they're all on. I'll connect you."

Novak heard a click on the speaker and took a deep breath. "Good morning, gentlemen. I apologize for moving this meeting up, but it seemed commensurate with the matter at hand. If we're all here as Cathy indicated, then for everyone on the line, we have General Forrester at Global Strike Command, General Cantwell at STRATCOM, Brigadier General Kean at OSI, excuse me, Dick, that's AFOSI, and also Lt. General Durwell at Strategic Deterrence over on the Air Staff. Gentlemen, please accept my appreciation for accommodating this call. I want you all to know that I have two Mossad agents in my office, Raphael Mahler and Meira Dantzig, and both are cleared for Top Secret. The director of Mossad, Mr. Taavi Perutz, brought this issue to my attention and loaned us these two agents to assist as they can."

"Let us have it, Pete," said General Charles Durwell.

"Thanks, Chuck. You always do cut through the bullshit and go right to the chase." Novak laughed, and the others joined in.

"All right, the bottom-line is this... we have learned of a probable, not possible, but probable plot between the Russian mafia and Vevak, the Iranian Intelligence Service, to steal a Minuteman III nuclear warhead, a MIRV. We assume this warhead would be loaded onto a Safir missile. We also assume the intended target is Israel. This links well to the recent announcement by the Iranians to open Iran to inspectors and stop developing nuclear weapons-grade material."

"Holy shit! That's a helluva way to start the morning out here in sunny Omaha, Mr. Novak," General Cantwell blurted.

"I know, General. My question to you is, can it be done? By the way, call me Pete."

"Thanks. Well, Pete, I don't see how. As Wallace will tell you as the Global Strike Commander, he owns these nukes, we only have three missile bases left. They're all in the northern tier... Montana, North Dakota and Wyoming. It'd be hard as hell to even approach those bases without being seen. Right, Wallace?"

"That's right, Pete. This sounds pretty wild," General Forrester said.

"What about storage? You have any warheads in storage?" Novak asked.

"Yes, but they're all on those three bases. They're kept in igloos, ah, bunkers, and each one of them stored on a bus. So, they'd be dealing with the same perimeter security. We've never, ever had a single breach of those bases' perimeters. Frankly, I don't think it's possible."

"Okay, thanks, General. But if we are correct in our assumptions, the Russians and the Iranians certainly think it's possible."

Novak pressed. "General Kean?"

"Yes, sir?"

"Dick, would it be possible to add some beef to your counterintelligence support to those three bases? Perhaps set up some surveillance posts in the towns that are closest? With an increased Security Forces coverage, we..."

Kean interrupted. "Pete, we have limitations on us, you know, the executive order?"

"This is in direct support of national security. The DCI and I will back you up one hundred percent and then some. Can you get counterintelligence teams up to those bases as soon as possible?" Novak asked.

"In that case... absolutely. And thanks for covering my back. I'll also send three additional teams of agents with some special weapons firepower to support the Security Forces there."

"Superb. Thanks, Dick. Gentlemen, can we all agree to this... we go to a higher THREATCON at just those bases for the next thirty days? Then, we'll reassess. Acceptable? If so, let's close this call and get back to our work. Thanks for attending."

Everyone replied in the affirmative, said good-bye and signed off.

Novak began to rise from his chair and the intercom buzzed again almost immediately, and Novak bent over and pressed the button.

"Yes, Cathy?"

"It's General Durwell back on the phone. He asked if we could go secure again. I can connect you in the secure mode, sir."

"Sure, go ahead. Put him on, Cathy."

Another click. "Pete? Chuck Durwell here. I didn't want to mention this at the meeting. Can I speak to you alone?"

"Chuck, Raphael and Meira are cleared for Top Secret."

"This is well beyond TS. It's Special Compartmented Information, and I do mean compartmented, a highly restricted

Special Access Program. They don't have access to it and probably never could have."

"What is it, Chuck?"

"The program nickname is unclassified... ARCTIC LANCE. Remember that program, Pete?"

"Ah, it's been years. It was connected with North..."

"Ah, ah, ah... Pete!"

"Sorry." Novak looked at Raphael and Meira. "Well, Chuck, is this program still alive, still viable?"

"Yes, it is. Your guys, along with the FBI, did the initial counterintelligence risk assessment for it."

"Yeah, okay, but years ago. How far along is it?"

"I'll have to check on that for you, Pete. But I think only the last phase is left before full deployment. I'll have to get back to you on that. Anyway, that's another distinct possibility. I know, who would think it with the cloak of secrecy we've had on that program."

"So, ARCTIC LANCE. Where was this again, Chuck?"

"Pete, I..."

"Where is it, Chuck? Come on, spit it out!"

"Alaska."

"Shit. Dammit. Taavi Perutz intimated they were looking in that direction." Novak grimaced. "Look, thanks very much. I'll look forward to hearing from you, soonest, please."

"Wait, Pete. One question... how do you think the Russians may have gotten wind of this to begin with?"

"A highly classified program that's years in development, such a protracted period of time... it's possible that sooner or later someone speaks out of turn and in the wrong place. Or a leak. I'm going to worry about that later, Chuck."

"Okay, thanks, take care."

"You too." The phones clicked off. The faces around Novak's desk had turned solemn.

CHAPTER 25

Alaska

The grizzled silver tips of the bear's cinnamon fur glistened in the pale light. The scent he followed was a confusing mix of fox and rabbit, but Brutus wasn't too particular about his meat, especially in winter. Plodding south, the grizzly drooled in anticipation, baring his jagged, two-inch-long incisors. On all fours, the massive predator was five feet high at his brawny front shoulders.

He plunged effortlessly through stands of birch and aspen saplings, the slender trunks frozen, snapping in his wake. Brutus lifted his nose into the breeze drifting up from the bay and grunted. The scent of blood in the air was strong. He was close. The grizzly quickened his pace, his four inch claws digging into the crusty snow, paws pounding the ground at a gallop.

At 4:00 AM, Brutus stood on all fours at the northern crest of the ravine, peering down through the dark. His head rocked back and forth uncertainly to sample the strange combination of smells that assaulted his nostrils. The wood smoke caused him to hesitate. On most occasions, that would mean fire and danger, but the enticing smell of blood was dead ahead. His ravenous appetite soon overcame apprehension. The bear stepped cautiously down the incline and entered the gulch, pausing occasionally to sniff the air. He passed by the first two tents, ignoring the sounds inside of heavy breathing and snoring. He was familiar with these strange creatures. Humans.

As he passed the entrance flap to the second tent, he followed his nose left, turning abruptly. Strong odors of fresh blood, meat, entrails. Irresistible. Brutus suddenly lunged at the red fox, snapping it into his jaws. The bear woofed, ripping through muscles and crunching the delicate bones. In minutes the carcass was gone, serving only to whet his appetite. Eager for more, the grizzly stretched his neck and sniffed higher, then stood up on his hind legs and moved closer to the tent's fabric wall. Rabbit carcasses, entrails and more blood. His claws snagged one of the rabbits as the other claw came down on the fabric. The tent collapsed under the crushing weight, its side wall splitting open.

Just inside the tent wall, Ivan Sokolov's screams ripped through the night as the grizzly crashed down across him, his cot's wooden frame splintering. Fumbling frantically for his pistol, Ivan fired round after round of 9mm jacketed hollow points into the bear's abdomen. Angered by the pistol's loud blast and sudden pain, the grizzly's ear-splitting roar echoed through the camp. His gaping jaws crunched down on Sokolov's upper arm, ripping it out of its socket and hurling it across the tent. Blood spurted from the gaping hole in Sokolov's shoulder as the Russian shrieked in agony and terror.

The bear dropped his full fifteen hundred pounds onto Sokolov's chest. Razor edged incisors tore into the man's throat, cleaving his neck down to the spine. With another bellowing roar, Brutus flung Sokolov's now-limp body into the air. It landed in a blood soaked heap at the foot of Nicolai Zaretsky's cot, who was cringing against the far side of the tent, praying that this was all some horrific nightmare. Snapping out of his stupor, Zaretsky yanked his own pistol from its holster and fired repeatedly at the huge hulk shredding the tent around him. Enraged at this new attack, Brutus slashed his claws across Zaretsky's chest, sending the pistol flying. Zaretsky crumpled to the ground, gripping his chest and thrashing in anguish.

Frenzied shouts reverberated throughout the encampment. Russians ran out from their tents half asleep in helter-skelter bewilderment, pistols and submachine guns in their hands. They converged on the crumpled tent just as Brutus leaped from the tent's flattened remains and sprinted north through the camp. Most of the men jumped out of the grizzly's path, but Junior Lieutenant Lipovsky stood his ground, raising his rifle. Before he could fire, the bear charged, trampling over him into the snow. The young lieutenant cried out in pain. The high pitched snap of rifle shots filled the ravine.

Snarling ferociously, Brutus dashed up the incline toward the north entrance. Major Krasnoff and Captain Ivanov jumped out from their tent together and collided instantly with the bear's monstrous frame, bouncing off the retreating grizzly. They gasped for air, the wind knocked out of their lungs, and fell back against the tent.

In the most fearful moment of his twenty-year life, Brutus bolted north through the darkness of the ravine and escaped up the trail to the woods. The rattle of gun fire continued, rounds spitting out and impacting the trees and snow behind him. The grizzly had been hit by no less than fifteen rounds, although none of the 9mm bullets posed a serious, mortal threat. Bleeding and in pain, he slowed to a lope and then a lumbering walk, wincing from wounds in his abdomen and shoulder. Dripping blood, Brutus retraced his previous path north through aspens and birches to his den where he could finally rest and lick his wounds.

The mayhem lasted only five minutes, if that, but it had wreaked utter havoc on the Russian camp. In their bare feet and underwear, the snow men milled around in shock, shivering in the frigid morning air. Standing in front of their tent and rubbing their bruised ribs, Krasnoff and Ivanov peered through the darkness at the stumbling shadows of their men. Ivanov was the first to recover, stepping forward, cupping his hands around his mouth.

"Men, go back into your tents immediately! Get your clothes on before you succumb to hypothermia." Hurry before you start dropping like flies!" Captain Ivanov shouted. "Don't come out until you have your full winter gear on. The first ones out, go to Sokolov and Zaretsky's tent... see what can be done."

The men disappeared into their tents, muttering to themselves without making much sense. Krasnoff slipped back into the tent, sat on his cot and began dressing. Before lifting the tent flap, Ivanov spun around again to the camp's center.

"Lieutenant Mikulich! Kolenka, you need to bring the emergency medical kit down to Sokolov's tent. There may be serious injuries. See if Sokolov and Zaretsky are okay, and start patching up whoever needs it. We'll be there in a minute," Ivanov shouted.

"Yes, Captain!"

Minutes later, the men gathered around the collapsed tent. Blood was everywhere. Sokolov's body lay in a tattered heap. He'd been nearly decapitated. His severed arm hung down from Zaretsky's cot.

"My God," Ivanov gasped. Krasnoff stood behind him, paralyzed at the sight.

Mikulich looked up. "Captain, I'm cleaning up Lt. Zaretsky's chest. He's going to need stitches. I have some local anesthetic that might help. Nothing punctured his ribcage though; should be okay. I'll need some light... somebody bring me a lantern. The sun won't be up for hours. Ah, Lt. Lipovsky... is just badly bruised. I'll give him something for the pain, but he'll be okay,"

Lt. Mikulich reported.

"Good work, Kolenka. We're lucky to have you."

Ivanov turned to the others. "Lieutenants Sorokin and Bardinoff, I know this will be difficult, but you two wrap up Sokolov's body and his arm in what's left of the tent and tie it securely. Place him behind the supply tent for now."

Ivanov paused, glancing up at the night sky. "Lt. Lipovsky, I know you're hurting, but get a fire going and heat some water. Get some pain meds from Kolenka first. It's not even 5:00 AM, but we'll all need some coffee and breakfast as soon as possible. We have a lot to clean up here. Oh, and somebody move Nicolai's cot and belongings over to Lipovsky and Bardinoff's tent. He'll have to bunk with them now."

"Yes, sir," Sorokin, Bardinoff and Lipovsky responded, their captain's brisk orders snapping them back to reality.

The men scattered to their tasks as Mikulich injected the anesthetic into both of Zaretsky's shoulders and began stitching. Mikulich nodded, grateful for the light as Lipovsky set a camp lantern down next him. Krasnoff gave the camp a 360-degree sweep of his eyes. Although severe, the disaster had been fairly localized: one man dead and a tent destroyed. He walked over to Ivanov.

"This was a tragedy, but you handled yourself well, Alexi."

"Thanks, Major. I'm going to go help the men with this mess." He turned his back on the major, and Krasnoff went back to their tent.

By 07:00 AM, it was still mostly dark with a faint light in the east. The Russians were finishing their coffee and nibbling eggs and sausages, their appetites were blunted by the morning's trauma. Conversation was sparse and muted. Afterward, they all quietly cleaned the cups and plates, then moved off to finish the work of getting the camp back in order and packed up.

As the sky lightened, the disarray became more visible. Lt.

Bardinoff could see more clearly as he cleaned up around the remains of what had been a tent. He knelt on the south side, now just tattered remnants.

"Major, Captain, please come look at this!"

Krasnoff and Ivanov hurried over to where Bardinoff was squatting. Scattered on the ground were the shredded remains of a red fox. The animal had been mostly ripped apart and eaten, but there was a razor sharp incision still visible on a residual section of its belly. It had obviously been sliced open. Not three feet from the fox was a rabbit carcass which, other than obviously being gutted with a sharp edge, was untouched. Their eyebrows rose in surprise and the snow men exchanged a grim look. The grizzly's attack on the camp was clearly no freakish accident. They'd been set up.

"Shit. Who would do this?" Ivanov growled.

"That fucking American agent. The one the Indian told us about on the bridge. The agent knew about this bear and sliced up these carcasses to draw it here. Dammit!" Krasnoff exclaimed.

Captain Ivanov's eyes steeled, drilling into Krasnoff. "Major, first we killed an innocent hunter, then we slaughtered an Indian and his young son, all noncombatants. The American agent must know this. Otherwise, he might have left us alone or never even found us."

Krasnoff turned to the men clustered around them. "Men, leave us now. Now!" As the Russians walked away muttering, Krasnoff turned to Ivanov, gritting his teeth.

"You're questioning my leadership, Captain? In front of the men? I am in command here, not you! I thought I made that clear at the start. It must never happen. Never. This mission is too important. If you do this again, Captain, I'll shoot you on the spot for jeopardizing this operation! Is that fucking clear?" Krasnoff screeched.

"Yes, sir. I apologize." Scowling, Ivanov bit his tongue.

"All right. We'll see."

Krasnoff swung about to face the camp center and shouted. "Men, all of you, listen up! Effective immediately, I want perimeter guards day and night at both the north and south ends of this ravine, fully armed with pistols and assault rifles. Captain Ivanov will assign the first shift. You will be relieved every six hours. Is that clear, Captain?"

"Yes, Major." Ivanov answered meekly, having been thoroughly embarrassed before his men.

"Lt., Sorokin, do you hear me?"

"Yes, Major?"

"You can operate the transmitter, yes?"

"Yes, sir."

"Good. I want you and Lt. Lipovsky to take the transmitter down to the shore. Raise the sub. Advise the commander of the bear attack and ask him if we can transfer the remains of Lt. Sokolov to the sub for formal burial at sea. Now. Hurry."

"Yes, sir."

The two lieutenants jogged over to the equipment tent, but were surprised not to see the transmitter in its usual place. After thoroughly searching the tent and boxes with no success, Lt. Sorokin stepped outside.

"Major Krasnoff?"

"Yes, what is it?" Krasnoff turned to them, annoyed at the delay.

"The transmitter… it's gone, sir. It's not in the tent."

"What the…?" The major and the captain ran down to the tent, but the transmitter was nowhere to be found.

"Sir? What now? How do we communicate with the sub?" Captain Ivanov asked.

His face contorted with anger, Krasnoff spun around. "We can't. I guarantee you that transmitter is somewhere at the bottom of the bay. And, this fucking American agent knows what he's doing."

"We should hunt his ass down and kill him, Major."

"Alexi, we don't have the time. And we need every man we have to execute the operation. That Indian said the American was well armed, but out here he can't communicate with anyone either. I'm hoping that having perimeter guards will keep this asshole away until we can move north. Hopefully, tomorrow."

"Well, I hope that too, Major, but I would like to see this sonofabitch dead."

"So would I, Alexi. So would I. We're down to seven men now. We just don't have the time to waste out in the wilderness hunting that clever bastard. Tell Lt. Sorokin to take Lt. Lipovsky to the shore and check on the rafts as well. If I am correct about this American agent, we don't have rafts anymore either."

"Sir? Yes, I'll tell them. But... our rafts? If you are right, how the hell will we get to the sub? How will we get home?"

"The mission, Captain Ivanov. The mission and only the mission is what matters. It's what we must focus on. You are Spetsnaz... you more than anyone else should know that. This is for Russia."

"And Sokolov?"

"You will bury him here, Captain. I suggest you get on it. We will have to find our own way out after the mission is complete. We can't burden that effort with Sokolov's body."

His shoulders slumped, Ivanov turned and walked away.

CHAPTER 26

McLean, VA

Pete Novak stood looking out his office window at the green hills of Northern Virginia. In a fleeting moment of whimsy, he wondered when the Autumn frosts would finally set the hills on fire with color. He walked across the room and then back again to the window. He'd skipped lunch. His appetite had left him.

Cathy stood in the doorway. "Sir?"

"Yes, Cathy?" He shook his head to dispel the daydream.

"Sir, you've been walking back and forth and staring out that window all afternoon. Is there anything I can do to help?"

"Ah... no, I'm just distracted, I guess. Anything else going on?"

"Yes, sir. The DCI... Mr. Barrett's on line 1 for you."

"Please put him through, Cathy." Novak walked to his desk and sat.

He picked up on the first ring. "Yes, sir?"

"Pete, this is Jack. Anything new?"

"Hey, it isn't often that the Director of Central Intelligence calls his Deputy for Operations. Usually, you don't remotely want to know what I'm into and if you do, you want a plausible denial."

Barrett laughed. "Well, I guess this is different. So, nothing?"

"I'm expecting some news, hopefully by close of business. Jack, I may have to leave rather quickly, and take two tactical squads along with me. I'm not sure where yet."

"You're in charge, Pete. I'm with you 100% on this and you have complete authority. From what I've been briefed, I can understand being puzzled. You don't actually have a wealth of information on this, just suspicions. Have you been to Delphi?"

"Thanks, Jack. Delphi? No, I hadn't thought of it."

"Hadn't thought of it or didn't see how it could help? You were never a lover of interactive databases. You always said the corruption of one poisons the whole pot. Right? Pete, my experience is that sometimes a trip to Delphi is worth it, and not just philosophically speaking. Sometimes it bears fruit, sometimes it's way too abstract in its answers. It can't hurt to check it out though. That's why you and I have terminals."

"You're right, I will."

"Just keep me in the loop, Pete."

"Yes, sir." The DCI clicked off.

Pete turned away from his desktop computer, swiveling his chair to the bureau behind him where the Delphi notebook computer sat. He flipped up the screen and clicked on the icon of a small glowing light in the upper left corner of the screen, smiling at the symbolism. The software engineer at NSA who built the program felt the glowing light represented the eternal flame at the temple of Apollo.

Delphi was a relational program that was all-source and all-database linked, including the FBI, ATF, DEA, NSA, NRO, NGIA, DIA, the military intelligence agencies, and seven foreign services as well as all intelligence reports received from CIA overseas stations. From up to ten words inputted, it examined relationships with the content of all known databases. Only a few directors and their deputies of national level agencies were accredited to use it, and the ability to speak at least two, preferably three, foreign languages was required.

Pete logged on as 'PNovak67!$' and entered a 15-character password: ***************

Delphi responded: *"Guten tag, Herr Novak. Heute ist Deutch."* "Hello, today is German."

Pete clicked on the window marked *"Bundespost."* "Post Office."

Pete entered five words: *"Krasnoff, Madani, Spetsnaz, Nuclear, Alaska."*

He hit the button *"Schicken."* "Send." And sat back to wait as Delphi searched all databases for any logical or suspected relationships. He turned and sipped the cold coffee on the desk behind him.

In just over five minutes, words flashed on the screen: *"Delphi. Sie haben post."* "You have mail."

Pete clicked on the mailbox, and again to open the envelope: *"Zug."* "Train."

"Train?" Novak eased back in his chair, staring at the word on the screen. He logged off Delphi, his fingers tapping on the edge of his desk. He pressed the intercom. "Cathy?"

"Yes, sir."

"See if you can get General Charles Durwell over on the Air Staff. He was on the conference call this morning. I'd like to talk with him, and I want to take this call secure again."

"Yes, sir."

Novak pushed himself up and walked to the window, but rushed back when the intercom buzzed. He plopped in his chair to respond. "General Durwell's on the line. I'll take you secure."

He heard the usual series of clicks, then Durwell's voice boomed. "Pete?"

"Hi Chuck. I'm glad I could get you."

"Well, truth is, I've been thinking about this whole thing too."

"Chuck, please just give me a brief overview... refresh my

memory on ARCTIC LANCE. I'm alone, the Israelis are elsewhere in the building."

"All right, sure. First of all, ARCTIC LANCE is basically about North Korea. At its inception, the program was predicated on CIA and DIA national intelligence estimates that a change in leadership in North Korea could exacerbate the situation there and send the country down the toilet. And, if it looked like the North was rapidly heading in that direction, the U.S. needed a means to stop it. At the time, we had none. Our assessment was that the entire Korean peninsula would turn into a moonscape. We couldn't let that happen. And hell, now look, we've had that change in leadership. You with me?"

"Yes, Chuck. I remember that now. Five years ago or more even, but I remember. And, so…?"

"Eielson Air Force Base is twenty-six miles southeast of Fairbanks. The Air Force owns several diesel locomotives at Eielson that are used to haul coal and other normal rail traffic, goods and so forth, in and out of the base. The Air Force and DoD decided that the ultimate solution for the Korean situation was to build another Minuteman III ICBM missile field at Eielson. Missiles launched there would have extremely short flight times, just minutes, really, with nuclear warheads in route to targets in North Korea. The CEP, ah, Circular Error Probability, for those warheads is about ten feet, so the kill ratio would be very high with such exceedingly precise strikes. And they'd be airbursts, so the ground would be recoverable at some point for repopulation."

"Go on, please."

"So, then the Air Force shipped one of its humongous tunnel-boring machines in from the Los Alamos National Laboratory in New Mexico. Those babies are huge. They build tunnels forty feet wide and fuse the walls in the process. The whole transit setup was staged from Idaho and went by rail up through Alberta, British Columbia, the Yukon and into Alaska and Eielson. A counterintelligence risk assessment on the entire rail route was conducted by FBI types. The construction operation itself… the whole thing really, the command module, silos and launchers, everything, was built from underground on up. Whenever required, things were moved about the missile field at night." He paused.

"And, all of it is camouflaged or tucked under hangars. All the equipment hauling just looked like normal rail traffic in and out of Eielson, in specially built boxcars that even accommodated the missile stages. That way, nothing would be visible from overhead, hostile nation space assets, until the system was ready to go fully operational. So, if the balloon ever did go up and North Korea looked like she was going to pop and head south across the DMZ, we'd take their collective asses out pronto with ten missiles and thirty thermonuclear warheads. End of problem."

"Good, I've got it. Chuck, then, the last question is, what phase, ah, where are we now on ARCTIC LANCE?"

"Well, last I knew, and that was six months ago, the only thing remaining was to bring in the warheads, again, by train. I'm waiting to hear on that. The whole damn program is so access-restricted, SCI compartmented and more. I can certainly understand that. And, due to the level of secrecy involved in such an operation, EPA doesn't know anything about this and we're not going to let some environmental impact study blow the cover on this. But I should hear soon. As soon as I know, I'll get back to you, Pete."

"All right. Chuck, thank you so much. It all comes back to me now. I thought the damn idea was dead."

"So did a lot of people. In a way, that's been part of the beauty of it. It's not on anybody's screen right now. Total secrecy."

"Chuck, do you have access to Delphi?"

"Delphi? What's that?"

"Ah, never mind. Don't worry about it. It's not relevant, really. Chuck, before we hang up, I need a big favor from you."

"And that is?"

"Please contact Alaskan Command on our behalf, Eleventh Air Force at Elmendorf Air Force Base in Anchorage. I need billeting arrangements in their VOQ for 23 people, including me plus a couple extra rooms for air crew."

"Okay, I'm following you. That should be doable. I know Lt. General Russell Holstrom there. He and I went to the Air War College together. So, beginning when and for how long?"

"Beginning tonight, Chuck. Let's shoot for a two week requirement for starters. And hey, I don't want my tactical response squads mixed in with the general population there, so try to get us a building, okay?"

"I'll emphasize to Russell that the CIA DepOps will be there and the whole thing is urgent, hush-hush and impacts the national security. Anything else?"

"Only that we'll need a hangar for one of our birds."

"Got it, Pete. I'll take care of it."

"Chuck, thanks for everything. Seriously, I owe you big time."

"Not at all. Take care and I'll call you when I get the latest word on the status of ARCTIC LANCE."

"All the best, Chuck. Novak out." The line went silent. Novak hit the intercom again.

"Cathy?"

"Yes, sir?"

"Get hold of Joel Butler for me and come in here together, please." Joel Butler was Novak's seldom used Executive Officer. Cathy was just so damned good that Novak rarely felt the need to turn to Joel. "Oh, and in the meantime, would you please grab me a BLT and chips from the cafeteria? My appetite's back."

"Yes, sir. I'll page Joel. The sandwich will be forthcoming."

Novak called the DCI. Barrett answered straightaway. "Yes, Pete?"

"Jack, I think I've got what I need. I'm going to Alaska tonight, taking two tactical squads and the two Israelis. We'll stage out of Elmendorf Air Force Base at Anchorage. I wanted you to know."

"Great, thanks. Keep me posted, please. If you run across and engage a Spetsnaz unit on U.S. soil…"

"Take them prisoner?"

"Hell no. Not unless absolutely necessary. We'll send their asses back to Moscow in body bags. Such an incursion would be unacceptable, a violation of international law and an act of aggression against the United States. I'll take care of explaining it all to the president. Any arrangements you need, make them in my name. Any challenges, have them call me personally."

"Yes, sir. Thanks very much for the top cover."

"You take care, Pete." Barrett signed off.

The BLT and chips were on Novak's desk in seven minutes. Cathy and Joel were sitting in front of him in eight. Novak had taken two bites.

"Thanks for coming. I need a couple of things accomplished. The DCI said to use his name if you have to, and Joel, you will probably have to."

"Yes, sir," Joe replied promptly.

"First, Cathy, I need a jet and crew to take us to Anchorage, Alaska, tonight. Elmendorf Air Force Base. It'll be me, the two Mossad agents, and two mobile, tactical response squads, twenty-three total. So, we'll need one of the larger jets in the fleet. Please call John Corrigan in Spec Ops and tell him I need two squads tonight, 6:00 PM. Prepare to stay up there at least two weeks. I want full tactical gear, vests, automatic weapons, the works."

"Will do, sir."

"Then, please call Walter Mackenzie and Bill Farris. Tell them to have Mr. Mahler and Ms. Dantzig at the hangar at Dulles at the same time, six o'clock. Raphael and Meira should pack up and check out of the Ritz Carlton; I'll meet them at Dulles. Just tell them that it's what they originally thought, no more than that. I think that's it for you, Cathy."

"Yes, sir." Cathy rose from her seat and left the office.

"Joel, I want you to call Fort Wainright, Task Force 49. It's just outside Fairbanks. Get hold of Brigadier General Shelton Atkinson. Use the DCI's name and mine. Tell him it's urgent, a national security issue, and will involve special operations but with our own Agency people. Tell him I need three UH-60 Black Hawks on the runway at Elmendorf Air Force Base no later than midnight tonight. I'll need them for ten days. Ask him to please have them equipped to transport 23 people total. Also, we'll need door gunners and M60D machine guns slung in the doors. The crews will have billeting at Elmendorf. Tell him the DCI and I greatly appreciate the support. Got it?"

"Door gunners and M60D machine guns? Got it, sir. This is going to be big, isn't it?"

"It's looking like that. If we have to engage, it may be one helluva gun battle, assholes and elbows, Joel."

"Good luck, sir. I'm on it."

Joel jogged out the door. Novak looked down at his BLT and devoured it in just a few bites, crunching the chips just as fast. Taking a long swig of cold coffee, he looked up and saw Cathy in the doorway

"Yes, ma'am?"

"Sir, Mr. Mahler and Ms. Dantzig would like to see you. May I show them in?"

"Sure."

Raphael and Meira walked in, concern on their faces. "Mr. Novak, do you have a minute?" Raphael asked.

"Absolutely. Please come and take a seat. Did you get the word?"

"Yes, sir, you want us to pack up and check out of the hotel."

"Yes, we're going to Alaska. We'll be flying into Elmendorf Air Force Base just outside Anchorage. If you don't have parkas or winter gear like boots or gloves, don't worry about it. We'll rustle them up for you. There'll be all the tactical gear we might need. Oh, and don't worry about the hotel bill. That's courtesy of the United States government for your support."

"Thanks very much," Raphael said.

"So you think it's Alaska after all?" Meira's brow furrowed.

"Everything points to it. A best guess. If anything happens, we'll be staging right out of Elmendorf, not from seven hours flying time away in McLean. We'll have two Agency tactical response teams, full gear, vests, full auto weapons, etc. For you guys too, if you'd like vests and more firepower. And three UH-60 Black Hawks are on their way to Elmendorf tonight. They'll be on the runway waiting for us whenever we need them, door gunners and all that. These birds can push 180 mph when they have to. We'll be ready. I'm just waiting on word from a contact in the Pentagon. So go pack up. Walter Mackenzie and Bill Farris will take you to the hotel and then the airport when you're ready We'll all have a late dinner on the plane."

"Mr. Novak, this is going full bore?"

"I think that's a good descriptor for it, Raphael… full bore. I've never been in favor of half-measures. We're making a very firm statement about the territorial integrity of the United States. Our tactical teams would call it 'assholes and elbows'. And, if it weren't for you two and Mossad, we'd still be totally in the dark about the whole damn plot. We're in your debt on that. So, thank you. I'll see you both at the airport. Shalom."

"Shalom," Raphael and Meira said in unison, smiling.

CHAPTER 27

Alaska

Gray sparrows hopped around in the snow just outside the tent, chirping and pecking at the brittle tops of dry grasses. McClure arched his head back to take a peek through the tent flap. One sparrow two feet away tilted its head to take a look at the large eyes peering out and abruptly flew away. He looked at his watch… 08:00 AM. It was full light out, unusually so for that hour. He'd caught some badly needed zzzs. It felt good, however brief. The night had been exceedingly stressful. He reached over into the parka and pulled out a couple of the Russians' food packs. All had photo covers, one looking a lot like beef stew and one like instant coffee.

McClure unzipped the tent flap, and the birds took off in a flutter of wings. The sun was up and radiating intensely off the snow. The temperature seemed to be well above freezing, nice for this early in the day, a promise of relative warmth. He wasn't in the mood for gathering kindling and digging a fire pit, nor did he have the time, so he pulled the sterno unit, a small pot and cup from the pack. He'd dine in this morning.

He scooped some snow from outside into the pot and set it on the lit sterno. Before long, he was devouring the beef stew and sipping coffee. To his surprise, the Russian stew was actually far better than the stuff he'd brought with him. He slipped on his boots, threw the parka over his shoulders and crawled outside to clean the cup and pot as usual, covering the spot with fresh snow to hide the smell.

As he stood, his eyes swept around his new camp, if he could call it that. Things looked much different in the daylight than they had last night. Aspen trees squeezed in on him. It really wasn't much of a clearing at all, just enough room for the tent. A narrow strip of space in front offered the possibility of a fire pit. No spruce bows padded the tent floor. And, he had no supply of kindling or firewood. Those things didn't bother him last night, but he'd tend to them when he returned. First, he had to see if his plan had succeeded, if Bruno or Brutus had paid a visit to the snow men.

McClure shoved his arms into the parka, pulled his pistol from the tent and holstered it. He set the rifle against an aspen trunk. Rummaging through his pockets, he found some crackers he'd scooped up in the Russian equipment tent. He'd take those along, and two Hershey bars. The water bottle was stuffed with snow. What else? Thinking for a moment, he decided to take one of his wire loop snares.

He wasn't going back to his first observation point, thinking that might be a little risky. If they were looking, the Russians could have easily found his tracks by now. This time he would retrace his arcing trail from the early morning, approaching the camp from the northwest, parallel to the shoreline. He fastened the snowshoes on and with the Weatherby slung over his right shoulder, swung his trekking sticks in front and tramped out from the aspens.

Following his previous tracks southward, he was immediately engaged with the splendor of the forest. The temperature had to be near 40 degrees, the air thoroughly stimulating. Even the rabbit population reveled in the magnificent day, scurrying all through the woods. He established a comfortable pace, the sticks swinging back and forth as quietly as he could manage. The snow was soft, the sun melting the surface cover. It was 10:00 AM when the bay came into sight, the water glittering in the brilliant sunshine. In fifteen minutes, his gait slowed as the southern entrance to the ravine came into view.

McClure hunched over and sank to his knees. Even with his polarized goggles, in the radiant blur of dazzling sunlight, he almost missed the figure leaning against a birch trunk near where the rafts had been. He was looking out over the bay, a submachine gun slung over his right shoulder. McClure gradually backed up until he could stand and move behind a small grouping of black spruce. He pulled out his binoculars and scanned the area. Only the one Russian, no doubt a perimeter guard. Dave pushed himself to his feet and reversed direction, quietly stepping away until he was entirely back in the woods and out of sight.

Something must have happened to prompt the snow men to post guards. He wanted to take a look. McClure trudged back north through the woods, eventually veering east. He figured he would now be approximately across the ravine from his old observation point. The forest was quiet, but he could hear a lot of chatter coming from the ravine. He pulled out the binoculars again and scanned the area left and right. Nothing. No guards here, most probably just at the north and south entrances to the ravine. He sat in the moist snow and gently unstrapped the snowshoes, leaving them behind with the sticks and rifle. With binoculars in hand, he squirmed his way to the brink of the ravine. He had no protective spruce on this side, but some brush grew along the ridge.

With the binoculars, he focused on the camp's center. It was buzzing with frenetic activity. The most southern tent on the east side where he'd placed the fox and rabbit carcasses was gone, just some scraps of torn fabric were scattered on the ground. Expansive smears of red blood stained the snow where the tent had stood. What might have been a shattered cot had been pushed aside, and the snow men were hurriedly moving boxes into the tent across the way.

It was apparent that some violent, destructive force had rampaged through the camp. Just as McClure had hoped, the grizzly must have been drawn into the ravine seeking food, and then responded instinctively when the snow men attacked him. The sheer amount of blood in that spot suggested at least some serious injuries. He didn't see a bear carcass anywhere, so after the confrontation, the grizzly must have managed to escape. McClure smiled grimly. These assholes had gotten a taste of what it was like to be on the receiving end. And, he wasn't nearly done yet.

He wiggled his way back from the edge, refastened the snowshoes and gathered his gear. McClure stared into the woods for a moment, thinking, and then took off to the southwest. In ten minutes, he was back in sight of the bay where he turned north, scanning for just the right spot. In just minutes, he found a thick stand of spruce ideal for the next step. Setting his rifle down, he pulled out his knife and started the search for some sturdy birch or aspen saplings. He found some birch just the right size not thirty yards away and began cutting one thin trunk into four stakes, each just over a foot long. Another straight, slender trunk he trimmed down to about six feet long, perfect for hewing into a throwing spear.

Back in the spruce grove, McClure unfastened the snowshoes and sat in the snow with a satisfied grunt. Stripping the bark off the stakes down to the white wood, he sharpened one end of each stake to a needle point. Two black spruce provided the perfect set-up. There was just enough space between them facing generally toward the ravine's south entrance. Positioning the snare between the spruce, he set the stake, and pulled the wire loop out to a diameter of a foot and a half.

Stepping off a distance of four feet back from the trees, he dug down through the snow and into the tundra to create a narrow trench directly in line with the snare. Luckily, the snow was only a foot deep here. He secured each stake in the trench a half a foot apart, packed into the tundra with a good six inches protruding, and covered the tips of the stakes with a layer of snow. Lastly, a spruce bough served as a nice broom to lightly brush the surface of the snow, leaving no trace of the stakes underneath.

McClure stepped back, sipping some water and admiring his work. It was perfect. He chuckled, imagining himself a regular Jeremiah Johnson, a man of the wilderness prevailing against overwhelming odds. Ready to sit and rest, he found a sunny spot, gathering a few dry twigs along the way. They might come in handy tonight.

He munched on crackers and one of the Hersey bars, and washed them down with water, saving the last chocolate bar for later when he would need all the energy he could muster. If everything played out according to plan, he would have to do some serious sprinting. To his advantage, he had snowshoes, and the snow men did not. With the snow bikes though, the Russians probably thought they wouldn't need snowshoes. That omission would cost them.

Nothing left for McClure to do now but fashion the tip of the spear into a javelin point. It was 1:00 PM. While he whittled, Dave considered the plan. His first attack had been conducted in absentia through the surrogate of a grizzly, and had apparently worked superbly. He hoped the grizzly hadn't been injured, at least not seriously. That attack was for Jack Stroud. What he planned to do tonight would be far more personal, more up close, though he hoped just as deadly. This one was for Robert and John Ewan. The snow men were going to pay damned dearly for that.

With only an hour of sleep last night and the sun beating down, McClure's eyes drooped. It was the warmest day he'd experienced since arriving in Alaska and as wonderful as it felt, the warmth wasn't helping. Determined to stay alert, he unzipped his parka, pushed back the hood and slipped off his gloves. Cooling off helped, but he still found himself fighting the urge to snooze. His vision was blurring. The trees, the snow and the shimmering water of the bay slowly faded in and out. Frustrated with himself, he finally scooped up a handful of wet snow to wipe over his face, and followed up by splashing cold water in his eyes. The chill was just the bracer he needed, but he was dumfounded to realize it was now 3:30 PM. He must have dozed off after all. It would be dark soon.

As the sun descended below the bay's horizon, the temperatures quickly began their downhill run to below freezing. McClure shoved himself to his feet, grabbed his rifle and stepped back ten yards to the west, where he leaned the rifle against an aspen trunk. It would be easy to reach if he needed it. He had to admit he was growing a little nervous. As justified as this plan was, he was going to kill a man tonight. Despite all the tactical marksmanship, close-quarter combat and hand-to-hand defense training at both the CIA and FBI, in the last ten years he'd never actually drawn a weapon. On the other hand, he didn't think he'd forgotten any of those skills. Those weren't things you forget. Very soon however, he would have to prove that theory.

At 5:00 PM, McClure tore off the wrapper and wolfed down the last Hershey bar. He strapped on the snowshoes and rose to his feet, then pulled the hood back over his head and slipped on the gloves, but he left his parka open in case he had a sudden need for his pistol or knife. He picked up the wooden spear and the dry branches he'd gathered and took a last look at the placement of his rifle. He spun about and headed toward the shoreline and the ravine's southern entrance. The night had turned from dark gray to black. The stars were out, but no moon yet. That would be to his advantage. The onshore breeze was picking up, and the cool air refreshed him and cleared his mind.

In twenty minutes, McClure dropped to his knees and pulled out the binoculars. The light-gathering optics proved superb. Once again, he saw just the one perimeter guard, no one else in sight. He could hear the faint chatter of men up in the camp. After the encounter with the grizzly early this morning, McClure hoped the Russians wouldn't expect another strike to take place so soon. He quietly backed up behind a black spruce. Feeling well concealed, he snapped a twig.

Lt. Stephan Lipovsky abruptly spun around, raising the barrel of his submachine gun. Leaning forward, he searched the dark forest for any movement. There was none. He'd heard the snapping of frozen branches before, but was it really that cold already? His eyes again swept the woods and the shoreline to the north and south. Still nothing. He shrugged his shoulders. He was on edge, that's all... but even so, pulled back the slide and chambered a round. The metallic click resonated in the stillness. He stepped out of the shadows of the trees toward the rock-filled shore, but not even the tree branches stirred in the slight breeze.

As Lipovsky walked back toward the trees, another twig snapped. He slowly turned to face the forest, eyes probing, but couldn't localize the sound. North along the top of the ravine's ridgeline was nothing but scrub brush, no place to hide. Well, maybe it was getting that cold. Another snap. The lieutenant looked north up the ravine, debating, but quickly decided they'd think he was stupid for calling out to them now. After all, he really hadn't actually seen anything. The wilderness at night had just spooked him.

Lipovsky raised the rifle to his shoulder and sighted along it up and down the shore. Nothing there either. At that moment, McClure moved out from behind the black spruce, moved forward and heaved the spear with all his strength. As Lipovsky turned to step back into the shadows, the spear struck low, slamming into his right boot and piercing it. The spiked tip stabbed deep into his foot just below the ankle.

"Arghh...!" Lipovsky screamed. "You damned asshole!" He fell to his left knee and jerked the spear from his boot. The lieutenant staggered to his feet and glimpsed the faint outline of McClure's form retreating into the trees. He fired a short burst that went high through the tree branches. McClure jumped behind the black spruce, crouching and waiting.

"Fuck you!" McClure shouted at him, scooped up some snow, and tossed a snowball high in the air at the Russian, striking him in the shoulder.

"You're one dead American pig!" the Russian shrieked. He fired another burst from the submachine gun and scrambled clumsily up the slope. His boots sank into the snow as he clambered up the incline.

McClure waited until he was certain the soldier was committed, then took off northwest into the woods. Struggling through the deep snow, Lipovsky fired another burst that ripped through the trees. Without stopping, he dropped out the magazine, jammed in another, jerked back the slide and fired again. McClure, sinking only a few inches into the snow with his snowshoes, easily outpaced the Russian, although the tree trunks splintered just a foot above his head from the 9mm rounds spraying around him.

In a rage, Lipovsky sprinted through the snow. He could see McClure. He fired again, the barrel bobbing up and down as he strained toward the fleeing American. Just ahead, McClure saw the two black spruce. He leaped over the snare, brushing the tree branches aside and angling to the right to avoid the stakes.

Bullets snapped over his head.

In the Russian camp, Krasnoff and Ivanov jumped to their feet at the sudden staccato sound of automatic gunfire. Coffee splattered on the snow as they dropped their cups and raced to their tents for their submachine guns. The other Russians, now on their heels, raced with their leaders down the ravine, hearing burst after burst of gunfire. Ivanov shouted for Lipovsky to stop and stay where he was... they were on their way to help. At the south entrance, Ivanov saw the blood tipped spear, spots of blood in the snow around it. Following the tracks, the group swarmed up the slope en masse.

Lipovsky was out of ammo, but he could see McClure clearly now, barely fifteen yards ahead. In one smooth motion, he swung the submachine gun over his shoulder and drew his pistol, firing without hesitation at the American. One bullet ripped through the loose parka flapping around McClure's waist. Ignoring the excruciating pain in his ankle, Lipovsky, focused solely on McClure, raced ahead. He was so close he could feel it, he was going to catch this prick who'd speared his foot, and he was going to kill him. Pointing the pistol's muzzle at McClure, he ran between the two spruce.

"I've got you, you pig!" The lieutenant shrieked, just as his right foot snagged in the wire loop of the snare. The anger on his face paled to sudden fear when he felt the loop tighten around his leg. Lipovsky's knees buckled as he hurled instantly down to the snow, his pistol still pointing ahead.

"Arghh! Nooo!" Lipovsky screamed in agony. The first stake plunged into his lower abdomen and through the muscle wall. The second gored him in the stomach. The third punched through his solar plexus. Lipovsky shrieked. The fourth stake speared through the lower rib cage, slid off his sternum and penetrated deep, nicking his heart. His eyes open wide in horror, Lipovsky's arm dropped to the snow. He looked up at

McClure and tried to raise the pistol once more, then fell face down, sinking in the snow.

McClure jogged over to kneel next to Lipovsky's unmoving form. He pulled the submachine gun off the shoulder, disappointed to find the magazine was empty. A search of Lipovsky's pockets revealed no spare magazines either . The Russian had fired everything he had at him. His eyes scanned the dark woods to the east. He could hear the other snow men swarming toward him. He picked up the submachine gun and jogged northwest at a hasty pace, slowing only briefly to snatch his rifle and sling it over his shoulder. He raced along as fast as his snowshoes would take him, following his tracks back through the forest. The Spetsnaz group would soon discover that he had one of their submachine guns, though they wouldn't know he had no ammo for it. He seriously doubted they'd pursue him tonight.

Ivanov was at least twenty yards in front of the group, running hard. He stopped in his tracks when he saw Lipovsky's body stretched out, face down in the snow. Bright red blood seeped out into the snow on both sides of him. Ivanov stepped forward and, seeing the wire around Lipovsky's leg, squatted down to remove it. Shaking his head, he hunched over the corpse as Krasnoff and the rest gathered around. It took two men to pull Lipovsky off the stakes, and turn him over, revealing the four gaping holes in the lieutenant's torso.

"No, no, no...!" Lt. Vladislav Bardinoff fell to his knees next to the body. "Stephan, my brother..." He reached out and laid his palm on Lipovsky's forehead, his eyes wet. Composing himself, Bardinoff stood and turned to Ivanov.

"Captain, Stephan was Spetsnaz, our brother, our friend. He would have given his life for any one of us. Sir, in less than twenty-four hours, the American asshole has killed two of our own. We must hunt this sonofabitch down, cut out his heart and nail it to a tree! Please Captain, we can't let Lipovsky and Sokolov's bodies just lie here buried in the wilderness. We must avenge them!"

Before Ivanov could answer, Major Krasnoff spoke, loudly and resolutely. "Pick this man up and take him back to the camp. We will bury him next to Sokolov, who also gave his life for Russia. Lt. Bardinoff, I understand your grief, but our mission is paramount. We don't have the time or the manpower to hunt this American pig down, not now. We have a higher calling. From now forward, no perimeter guard will engage in a confrontation with this American or any other intruder. You will fire your weapon to alert us and call for help. Is that understood?"

"Yes, sir," the men replied in unison.

Lieutenants Bardinoff and Mikulich picked up Lipovsky's bloodied body and carried him eastward to the ravine. Captain Ivanov lagged behind the group, gritting his teeth, his face creased in anger. He was at the outer limits of his tolerance. He wouldn't take much more.

CHAPTER 28

Alaska

Lt. Vladislav Bardinoff stared up at the tent's ceiling gently rippling in the onshore breeze. It was 8:30 PM. Major Krasnoff had ordered them all to go to their tents and get some sleep. Tomorrow they would execute the final phase of the operation. At daybreak, they would break down the camp and all of them would then head north to the bridge. But Bardinoff couldn't sleep. His eyes wet, he glanced over at the empty cot where Stephan had slept. He burned both with sorrow for Lipovsky's death and a seething rage to avenge that death.

He knew for certain that Captain Ivanov wanted this American pig dead as well. The captain was with his men on that. Letting these two deaths go unavenged was not the Spetsnaz way, but Krasnoff couldn't understand that. Bardinoff knew in his heart that he could beat this American. He could pound his ass to a bleeding pulp and then cut out his heart. Tracking him through the snow would be easy even at night, he thought, turning restlessly on his cot.

The American would be off guard and tired, would never expect something tonight. Bardinoff eased off his cot and peered out through the tent flap. The fire in the pit was sputtering down. Nobody was out except the perimeter guards. Lt. Mikulich was stationed down at the south entrance. Kolenka would understand, he would surely let him pass, if he promised to be back before sunrise.

Bardinoff sat back down on the edge of his cot, thinking. He shook his head and rose abruptly, his dogged resolve forcing the decision. He slipped on his white snow pants and shirt and laced up his boots. The parka went on next, along with the wide leather belt that secured the pistol and the knife around his waist. He slipped out the mag… loaded.

Bardinoff then slung the submachine gun over his shoulder, stuffed two extra mags in a pocket, and cautiously edged through the tent's opening. He stood silently in place, listening. Not a sound. Everyone in the camp was either dead tired or profoundly stressed out or both.

Bardinoff gingerly tip-toed through the camp center, past the fire pit and beyond the large equipment tent. As quietly as he could, he made his way down the trail to the ravine's south entrance. Mikulich was leaning against a tree, facing out to the bay, puffing a smoke. Bardinoff spoke softly so as not to alarm him.

"Kolenka. It's me, Vladislav."

Mikulich spit out the cigarette and swiveled around, rifle at the ready. "What? What the hell are you doing up? Shift change isn't for hours. It's not time yet!" He lowered his weapon when Bardinoff stepped out of the shadows.

"Kolenka, I'm going to track that American bastard down and kill the fucker. Let me pass. I'll be back before sunrise, I promise," Bardinoff pleaded.

"Impossible, Valdislav. I can't let you do that. It's against orders. That jerk Krasnoff will hang me."

"No, he won't. He can't. He needs you. He needs every man he has, but especially you because you know that locomotive better than anyone."

"No…!"

"Kolenka, look, we have no transmitter and no rafts. We can't call the sub and we can't get out to it. After a period of no contact, you know, the sub commander will have to leave. They can't stay here forever. If I can find and kill this American, then he can't tell anyone else about us. Maybe then we can hike out of here, buy some clothes, get up near Nome. We'd at least have a chance. The American's alone. I can do this, I know it."

"No. Everything you say is true, Vladislav, but I can't let you do this." Mikulich raised the barrel of his submachine gun.

"Oh, so now you're going to shoot me, Kolenka? You'd kill a brother? Okay, well, let me make it easy for you." Bardinoff turned away from Mikulich and stepped toward the woods.

"Stop!" Mikulich pulled back the slide.

"Go ahead, Kolenka. I wonder what the others will say? Look, I'll be back before dawn and this asshole will be dead. I'll bring you back his heart. Take care, brother." Bardinoff rambled through the snow and up the slope, waving his hand behind him.

"Dammit," Mikulich muttered, lowering the barrel and watching as Bardinoff disappeared through the trees. He shouldered the sling of his weapon, pulled a pack of smokes from his jacket, stuffed one in his lips and lit it. "Dammit," he whispered again.

Following the tracks in the snow was child's play. Before he knew it, Bardinoff approached the two black spruce trees where Lipovsky had died. Large dark stains marked where his life's blood had drained out. Bardinoff stood over the spot.

"This is for you, Stephan. Sleep well, brother, and know that you are loved." Bardinoff looked at the tracks ahead and stepped off, trailing them through the forest. He slogged his way northwest, keeping his eyes on the American's path. The depth of the snow varied, which made the hike exhausting, draining him. In no time at all, his thighs ached for relief. After probably an hour, he realized the tracks had shifted to the north and somewhat east. He slowed his pace, concentrating now on moving as quietly as possible. However, the depth of the snow made stealth nearly impossible.

At quarter past 10:00 PM, McClure had already dug a fire pit, lit a small fire and cut an plenty of spruce boughs which he'd spread out to cushion the tent. There was even a half hour or so for a short snooze. Tomorrow, he planned to build another lean-to for the campsite. His supply of kindling was low, especially the larger branches. Without much thought, he dropped his rifle and pistol in the tent and set out to cut some additional wood, his mind still busy. He discounted worrying about bears. One grizzly was likely wounded, perhaps badly, and at the very least most certainly scared half to death. All the gunfire would most likely have driven the other grizzly far up to the north as well. So the bears really wouldn't be a concern for a couple of days. McClure looked at his load, realizing with a start that he had an armful of large branches. Laughing at his distraction, he circled back toward the campsite.

As Bardinoff crept along the path, he noticed the tracks were now spaced closer together. The American pig must have slowed his pace, he must be near. He pushed back his hood and slipped off the gloves, his eyes skimming the forest in front of him. A small yellowish flicker of light caught his attention, maybe a hundred yards away.

A campfire. He crouched low, slowly pulled the slide back to chamber a round, and switched the safety lever off. Hunched over and pausing often, he moved cautiously forward toward the camp, weaving his way around bushy undergrowth. The campfire was low, so the American might already be in the tent and asleep. Bardinoff smiled. He had the sonofabitch. A rabbit suddenly burst out from the bush at Bardinoff's feet. Startled by the movement, he backed up abruptly. "Dammit!" he muttered.

McClure heard the sharp snap. Instantly he froze, standing motionless, staring southeast. That was not the popping of a frozen branch. He squatted down noiselessly to his knees and slowly, laid the branches in the soft snow. Both his rifle and pistol were back in the tent. *Nice move, Dave. What was that you read about overconfidence in the wilderness?* he thought ruefully. Wiping the SOG titanium knife on his pants, he slid it gently into its sheath, then set his gloved hands softly on the snow to crouch on all fours. Barely breathing, he inched his way back behind a spruce, eyes studying the forest to the southeast.

Movement. McClure squinted in the moonless night to make sense of the form creeping low and slow through the stand of birches to the south. The watery gleam of light rising from the snow made it difficult to distinguish between the still-life shadows of trees and the skulking figure, hunched over and moving with great caution toward the campsite. From thirty yards away, the figure caught the faint illumination of the fire. A man with broad shoulders in a white parka and pants with black goggles pulled down over his eyes. One of the snow men!

McClure's stomach fluttered and a tingling rush swelled over him. Even as he felt the goose bumps rising, he recognized he had the advantage. The Russian, wholly focused on the camp, was distinctly visible to him. And, while glancing briefly right and left as he stole forward, he was evidently certain McClure was in the tent. That mistake would cost him.

McClure quietly slipped off his gloves and parka and set them on the snow, shivering in the icy air. He was freezing but he wanted as much freedom of movement as possible. He slowly drew the knife from its sheath and, crouching low, he began creeping south, parallel to the white-clad figure. As the Russian snuck north, McClure circled silently behind, intent on closing the gap. He was directly behind the Russian.

Fifteen yards from the tent, Bardinoff rose from his crouch

and eased forward one step at a time. The submachine gun balanced at his waist and aimed at the tent, his finger on the trigger. McClure held his breath, now just five yards behind. Bardinoff tilted his head, examining the tent. Curious – the flap was down but not zipped. He shook his head, shrugging, and pulled the trigger. A deafening burst of automatic gunfire exploded from the muzzle, spraying the tent with bullets, shredding the fabric. The earsplitting rattle of gunfire drowned out all other sound.

In the midst of the thunderous clatter, McClure vaulted forward, driving his shoulder into Bardinoff's back and thrusting the razor-edged titanium knife through the man's ribs. The blade pierced the right lung.

"Ahhh!" Bardinoff shrieked. The collision knocked him forward, flinging the submachine gun out of his hands and plunging his face down into the scorching embers of the fire. "Arghh!" he screeched at the blistering pain, clawing at his eyes.

McClure barreled over Bardinoff and rolled against the front of the tent, the knife still in his hand. A film of snow was pasted across his forehead. He instantly sprang to his feet.

Bardinoff scrambled out of the scalding fire, his face blackened and his right eye closed and seeping. He wheezed and gave a gurgling cough, blood spewing from his throat. He wiped his mouth, glowering.

"You're one dead bastard!" Bardinoff screamed in English, his right hand drawing his pistol.

McClure's eyebrows rose in shock. With no hesitation, he dove low, his shoulders smashing into the Russian's knees. Bardinoff yelled out in distress and fell forward over him, the pistol skittering into the brush.

McClure spun around and jumped at Bardinoff again, but the Russian raised his boots and slammed them into McClure's chest, stopping him in midair and dropping him to the snow, gasping. Bardinoff shoved to his knees, drew his knife and plunged it into McClure's left shoulder.

Despite the searing pain, McClure leaned forward into Bardinoff's knife thrust, driving the blade deeper into his own shoulder. Bardinoff realized too late why. Now much closer, McClure swung his right arm forward in an arc. The SOG knife blade caught the side of Bardinoff's neck and sliced inward. The Russian's eyes opened wide in horror. McClure pushed himself up, shoving the knife forward and slashing across the Russian's throat, ripping through the esophagus. Blood spurted across the snow from the severed carotid. Bardinoff slumped to the ground, his body jerking violently and his eyes wide in panic. In a moment, he lay motionless.

McClure fell forward and rolled off Bardinoff and onto his back, his lungs wheezing for air. Blood seeped from his wounded left shoulder. His knife, right hand and arm were drenched in blood. He lay still, chest heaving, staring up through the branches that seemed to grasp at the black sky. Every ounce of energy was drained from his body.

He looked in the direction from where Bardinoff had approached. There was no one else. Just this one. That was enough for one night. His head eased back onto the snow. A wave of emotion began to surge up in him. Tears welled up in McClure's eyes, and he jammed his knife into the ground. His fists pounded the snow, then slumped to his side.

"God... " he whispered into the frigid Alaskan night. "God, thank you for another chance."

He couldn't stop the tears. He faced up to the stars and cried. He was finally free of the guilt and grief clinging to his spirit for over a year.

CHAPTER 29

Alaska

An icy gray fog had crept in from the Bay of Alaska, blanketing Seward, Valdez and everything to the northeast. It hung suspended ten feet above the snowbound roads, towns and forests. Temperatures dropped to just below freezing. Branches of aspen, birch and spruce trees became coated with rime, as if dipped in sugar-frosting. Farther inland, the weighty fog dispersed, becoming a lighter mist.

Sometime during the night, McClure had found the energy to slip out of his boots and crawl into the bullet shredded tent. Even riddled with bullet holes, the sleeping bag still offered abundant warmth, and he immediately fell into a black, dreamless sleep. When he blinked his eyes open and lifted the tent flap, a clammy chill clung to his face like a damp washcloth. The blood-soaked body of the Russian lay just outside the tent, jolting him wide awake and reminding him of the sharp pain in his left shoulder. He reached up and probed the wound, sticky with blood and sensitive to even the slightest pressure. He needed to attend to that and now.

McClure looked at his watch... 08:30 AM. The sun was up but well hidden behind the low fog. He unzipped the sleeping bag and sat up, digging in a pocket on the front of the backpack for the little sewing kit. His hand scrounged around until he found the bottle of Jack Daniels, what remained of it. He unbuttoned his outer shirt and pulled down the neck of the tee shirt, flinching as the fabric pulled away from his crusted skin.

The blade had gone in nearly three inches, he guessed, feeling lucky. Clenching his teeth against the sting, he drizzled whiskey over the wound, scowling and thinking, *What a terrible waste of damn good whiskey.* He threaded the long needle with a length of the thickest thread in the kit and dripped whiskey over both. McClure flinched as the needle pierced the raw flesh. Steeling himself, he pushed the needle repeatedly through the skin, stitching the wound closed and double-knotting the end of the thread. He was surprised that the actual doing wasn't as bad as the anticipation of it.

More whiskey went over the stitching, and he took a swig himself. As he slipped the sewing kit back in the pocket, he pulled out his pill container and shook out three Motrin. The pain killer went down nicely with another swig of whiskey. He pulled his last clean tee shirt out of the pack, folding it into a square pad, and taping it over the sutured wound with adhesive from his emergency kit.

McClure took another swig of whiskey, buttoned the shirt and eased gingerly into his parka and boots. Unstrapping the pack shovel, he crept out from the tent and stood, staggering slightly. He felt lightheaded and heavily bruised from last night's death match. At his feet, Bardinoff was frozen solid, eyes open and face a pale blue. It took him over an hour, but he managed to drag the body some thirty yards from the tent and bury him in a shallow grave. He shoveled fresh snow on top to suppress any odor. Exhausted from the effort, he plunked down in front of the tent and heated up some chicken noodle soup and coffee on the sterno. He ate slowly, savoring the taste of the food. Again, the Russian food pack was much better his own. Now that the Motrin had kicked in, he felt better, more relaxed.

After cleaning up, McClure stuffed some crackers in his pocket and snow in the water bottle. Holstering the pistol, he packed the rifle's three extra magazines with three rounds each. He looked in the tent for the rifle, and saw splinters of wood around it. There were two 9mm bullets imbedded in the stock and one had cracked the forestock grip, but it held together. It was from the Russian's spray of machinegun fire. He jerked the bolt back and forth and found no problem, the rifle was operable. The scope and mounts were unscathed. Not a scratch. Very lucky. He pulled out his knife and gouged the bullets out of the wood.

Taking a quick glance around, the thought crossed his mind that he might never return to the camp. McClure strapped the snowshoes on, gingerly settled the rifle strap over his right shoulder and stepped out from the camp at a slow, but deliberate pace, heading north to the bridge. He knew he was in no condition for another fight, but he had to find some way to stop the snow men. He thought he might be able to find a concealed spot on the west side of the canyon. If he had enough time to pad a nice bench rest from the snow, he might be able to take them out one by one with the Weatherby. He had no other option; another hand-to-hand battle was out of the question. He'd never survive it.

While McClure was struggling to dig Bardinoff's grave, Major Krasnoff and Captain Ivanov picked their way through the ravine's fog-shrouded path toward the shore. They didn't like what Lt. Zaretsky, the relief perimeter guard for Lt. Mikulich, had jogged back up to the camp to tell them. Zaretsky silently trailed a few yards behind the two. When Mikulich saw the group approaching, he stepped away from the birch trunk against which he'd been leaning, threw the cigarette in the snow and crushed it with his boot.

"Lt. Mikulich, where is Lt. Vladislav Bardinoff?" Krasnoff

barked. Ivanov stood quietly next to him.

"I tried to stop him, major, but he turned his back on me and walked away." His eyes dropped. "I couldn't shoot him."

"Let me repeat the question, Lieutenant, where the hell is Lt. Bardinoff?"

"He left last night, sir. He said he was going to track the American down and kill him, that he'd be back before sunrise. But... I haven't seen him since."

Inches from the lieutenant's face, Krasnoff roared, "Bardinoff hasn't returned because the stupid shit is dead!" Pulling his pistol from his holster and ramming the muzzle against Mikulich's forehead, Krasnoff snapped his head sideways, freezing Ivanov's sudden movement with a single glare. Ivanov froze in place.

Krasnoff turned back to the lieutenant, shoving the pistol even harder. "What you've done here, Lt. Mikulich, is a court martial offense! If I didn't need you to drive that diesel locomotive today, I would this instant put a bullet through that worthless pound of protoplasm you call your brain! Do you understand?"

"Yes, sir." The lieutenant's face was stark white, his eyes bulging.

"This American agent is no dumb ass hunter. He is obviously highly skilled and he is clever. We're down to five men. Shit!"

Captain Ivanov stepped forward. "Mikulich, you and Lt. Zaretsky get back up to the camp and help break it down. Unpack the EMP rifles and unhook the battery waist packs and backpacks from the chargers, ensure they're fully charged, then turn off the generator. Go! Now!" Ivanov met Krasnoff's glare for a long moment, then turned on his heel and walked back up the trail.

By 10:00 AM, the five snow men had climbed on their three

snow bikes. Krasnoff sat behind Ivanov on one, ready to tow a heavy sled. The major was dressed in a U.S. Air Force cold-weather parka, lieutenant colonel's blue flight cap, blue pants and boots. Lieutenants Sorokin, Mikulich and Zaretsky wore the battery backpacks and waist packs, and carried the stocks of the long barreled EMP rifles across their thighs. Railroad track equipment was strapped to the sides and rear of their bikes. They had to hustle. Krasnoff shouted at them to hurry. The train carrying the warheads should have already left the depot at Haines Junction an hour ago and was due to cross the bridge between noon and 12:15 PM.

The wintry fog was dense as they rode north, trees frosted white looming like ghostly sentinels along their route. The freezing mist had left a thick coat of ice on the snow's surface, making the trail treacherous and slowing their progress. But as the group traveled farther north, the fog slowly dissipated, the sun broke through the clouds, and the snow on the trail softened, making the ride smoother and faster. The Russians finally pulled their bikes into the woods and off the narrow path along the canyon's eastern edge at 11:30 AM.

The three lieutenants set the rifles against the bikes, and scrambled to remove their gear. Tools in hand, they hurried to meet Ivanov waiting at the bridge. Krasnoff walked up to the track and turned east along the rail, checking the brush cover at the shooting positions they'd selected. All appeared adequate and in order. Ivanov paired with Sorokin and Mikulich with Zaretsky to remove the heavy hardware that secured the first section of rails on the bridge. Fifteen minutes slowly ticked by, Krasnoff watching them from the east edge of the bridge, constantly checking his watch and shaking his head at the slow progress.

Sweating from the frantic labor, the four men finally kicked the rails loose and ran off the bridge, tossing the track gear clanging down into the canyon. As they strapped on the battery packs and picked up their rifles, Krasnoff checked the time and nodded with relief: 12:05 PM. They'd made it.

Plugging the rifles into the battery packs, the three lieutenants walked down the tracks to their shooting positions, disappearing into the woods. Krasnoff and Ivanov pulled out their pistols and jerked back the slides. They nodded to each other, Ivanov gesturing a thumbs-up. The major stepped out into the middle of the track.

"Remember, when I wave down the train and walk alongside the locomotive, the engineer will open his window. All three of you are to shoot him directly in the head, several shots each, making sure you hit skin to allow the EMP pulses to do their work? It's critical that you incapacitate the engineer at the very beginning."

"Yes, sir," the lieutenants shouted.

"At that point, Lt. Mikulich, you immediately drop off your waist pack and back pack. Leave the rifle there. Captain Ivanov will assume your firing position. Run to the diesel locomotive, climb in and ensure the train is ready to roll. If the security car contacts you on the cab's intercom, tell them in your best English that the track on the bridge is being repaired and the work is nearly completed. Then, jump down and join me."

"Yes, sir!" Mikulich replied.

"Captain Ivanov, Lt. Sorokin and Lt. Zaretsky, if any of the security force attempts to exit their car, shoot them in the heads with the EMP rifles and keep firing until they drop. Remember, you each have about twenty-five shots until the batteries give out."

"Yes, sir!" They shouted.

"Lt. Mikulich, you will remove the fuel tank covers, then pull the sled up to the nuclear container car and assist me as I direct you. When I tell you to return to the locomotive, accelerate the train to the bridge. You'll have to jump off before the locomotive hits the loosened rails."

"Yes, sir!" Mikulich answered.

"Good. Lastly, and this is vital. The body of the engineer stays in the locomotive cab. Any bodies from the security force must be put back on the train, in the container car, if necessary. Is that clear?"

"Yes, sir!" In unison, they turned their heads to the east, hearing the muted roar of the train approaching. Krasnoff glanced at his watch… 12:10 PM, right on time.

Captain Ivanov shouted, "Remember, men, there will be almost instantaneous nuclear contamination when train crashes into the canyon. We need to distance ourselves from it quickly. So, head as fast as you can east down the track with your EMP rifles and battery packs. We'll dispose of them later somewhere in the forest along with the snow bikes. Then, we can sit down and determine our route out of here." He hoisted his rifle overhead and shouted, "*Uraa!*"

"*Uraa! Uraa!*" Major Krasnoff and the lieutenants cheered.

"Good luck, men!" Krasnoff shouted and stepped to the center of the track.

On the west side, McClure arrived at the canyon panting with exertion, the stab wound throbbing in his shoulder. He dropped to the snow, eased the rifle off his shoulder, and unstrapped the snowshoes, trying to catch his breath. On all fours, he snuck to the edge of the canyon, peering across to the east side of the bridge. He saw the snow bikes across the chasm, parked just inside the woods, but the Russians were nowhere in sight. To get a bead on them, he'd have to move further north, maybe all the way up to the tracks. His eyebrows shot up as the train's whistle echoed through the forest as it approached the bridge.

McClure jumped to his feet and raced north through the

woods as fast as his legs could go. The snow wasn't all that deep. He moved out from the dense tree cover toward the bridge and kneeled down in the snow to sight with the rifle scope. He had a good angle, seeing three of the five men right away. Major Krasnoff was standing in the middle of the track pulling red flags out from under his parka. Puzzled, McClure centered the crosshairs on Krasnoff for a better look. McClure could make out a U.S. Air Force officer's flight cap on his head. His brow furrowed, wondering what their game was, but couldn't dwell on it. He had to localize the two missing Russians before firing to be sure they weren't on his side of the bridge. His scope scanned the south edge of the woods along the track and spotted both of them. Good. Krasnoff walked east along the track, distant but still visible.

The train came into view as it rounded a curve in the track, slowing as it approached the bridge. Krasnoff raised the red flags over his head and waved them furiously back and forth. The train's whistle pierced the air again. The locomotive braked but not fully, the engineer obviously not happy about slowing his train. Krasnoff stood his ground, waving the red flags to signal danger. The engineer braked hard, the wheels grinding to a stop just ten feet from Krasnoff.

Krasnoff lowered his arms and walked toward the cab as the engineer stuck his head out the window and shouted, "What's going on? What the hell do you think you're…?"

Five bursts from the EMP rifles hit skin just below the engineer's left temple, causing instantaneous cardiac arrest. His eyes went blank and his head fell limp, banging down on the metal window sill then sliding backward and disappearing into the cab.

Mikulich dropped his batteries and handed the rifle to the waiting Ivanov. The lieutenant sprinted up to the train, bent over to release the diesel fuel tank caps and then scrambled up the ladder into the cab. In just minutes, he jumped out and ran wheezing toward Krasnoff. The security forces car was the last of the two cars trailing the engine, just behind the car with the warheads. It's doors suddenly slid open and two airmen jumped onto the track, rifles in hand. Both dropped immediately from the barrage of EMP shots, and someone inside promptly pushed the door shut, leaving a small crack of space just wide enough for a rifle barrel. They fired, spraying the area randomly. Krasnoff swore explosively when the snow at his feet was suddenly peppered with bullets.

The first .300 magnum round from the Weatherby rifle hit Mikulich as he ran, ramming into the right side of his neck. The force of the round abruptly hurtled him sideways as the expanding bullet blew through his neck, splintering the spinal cord and tearing open both the carotid artery and the jugular vein. The lieutenant fell dead at Krasnoff's feet. Krasnoff ducked instantly into a crouch, firing his pistol at McClure's position across the bridge. Hunched over, Krasnoff retreated to the weapons container car.

Lt. Sorokin leaped out of hiding onto the track, repeatedly firing into the woods toward McClure as Krasnoff shouted for Zaretsky to get the snow bike and assist him. Ivanov joined Sorokin and grabbed his arm, pointing directly at McClure across the canyon. Both dropped their EMP rifles and pulled out their pistols, spraying the area with 9mm rounds. McClure ducked his head as bullets slammed into the surrounding trees. He felt a tug in his right shoulder. The 9mm round had such high velocity that it didn't hurt at first, but seconds later a searing pain made him flinch.

"Dammit!" McClure screamed. He shoved to his knees and, ignoring the pain, pushed the rifle stock firmly against his burning right shoulder. The crosshairs of his scope on Sorokin's chest, McClure fired, briskly slammed the bolt back and forth to chamber another round and fired again. Both bullets thudded into Sorokin's chest, one ripping through a lung and the other rupturing a ventricle of his heart. Blood spurting from his chest, the pistol slipped from Sorokin's hands and clattered to the metal rails as he fell face down on the heavy wooden ties.

Ivanov reached for Sorokin, but had to dive for cover as another bullet snapped through the air past his head. He fired at McClure as he went down. Dodging and sidestepping, he dashed to the engine, sailed up the ladder and threw himself in the doorway. He found the accelerator lever and firewalled the engine. The locomotive jerked forward as Ivanov leaped from the moving platform and hit the ground rolling. Before he was fully on his feet, the pistol was already in his hand and aimed at the American.

McClure couldn't see Krasnoff. He pushed himself to his feet, dropping out the spent magazine, and popped in another as he sprinted to the center of the tracks where he dropped prone.

In the container car, Krasnoff fell to his knees as the train lurched forward. He rose quickly and pushed the container with the warhead out so Zaretsky could ease it onto the waiting sled. They strapped it securely in place. Krasnoff checked it, slapped Zaretsky on the back with a broad grin and jumped on the snow bike.

McClure thrust himself up to a standing position to get a clear shot. Immediately, six gun shots rang out from Ivanov's pistol. As McClure squeezed the trigger, one of Ivanov's bullets struck him in the left thigh and the Weatherby's muzzle jerked from the impact. The bullet plowed into Zaretsky's lower abdomen, exploding through his intestinal tract and out his right kidney. The lieutenant collapsed to the track curled in a fetal position, screaming and writhing in agony.

McClure grimaced at the new throbbing pain in his left

thigh, but managed to sweep the scope's parallax, scanning for Ivanov. He fired just as Ivanov dove for the track. The shot missed. In the frenzy of the battle, McClure lost track of his rounds. To be safe, he dropped out the magazine, shoved in another and chambered a round. Ivanov jumped to his feet and raced back to Krasnoff. McClure followed with the scope.

"Major Krasnoff! Major! Where the hell do you think you're going with that? You intend to kill more noncombatants, more innocents, you bastard! Halt!" he shrieked, raising his pistol, but Krasnoff was faster. Two shots blasted from his pistol, thudding into Ivanov's left chest and shoulder, and dropping him to his knees. Krasnoff gunned the engine, and the bike wailed as it jerked forward and sped east down the track.

McClure settled the Weatherby on his shoulder and flattened his jaw against the stock. Tracking the crosshairs off Ivanov and on to Krasnoff quickly disappearing down the track, he held his breath and gently squeezed the trigger. The .300 magnum round exploded from the muzzle on a flat trajectory toward the escaping Russian.

Krasnoff shrieked at the top of his lungs as the bullet slammed into his lower back, blew through his right kidney and burst out his stomach. Pitching forward on the seat and sprawling off balance, cracking his head on the gas tank cap, Krasnoff somehow managed to hang on to the handlebars. Blood flowing from his back and abdomen, he pushed himself up and swerved the bike off the tracks, into the woods and out of sight.

Frozen in shock, McClure watched the train hurtle toward him. It reached the bridge at 30 mph, enough speed to catapult the loose rails out from under the forward wheels and whip them end over end into the canyon. Wooden ties spewed from underneath as the locomotive's momentum drove it forward. The engine upended, teetering, then plunged off the canyon. The cars followed, ripping apart as they careened down the walls of the canyon.

What was left of the security force never had any hope of escape. The diesel fuel tanks ruptured, a muffled detonation that sent a cloud of fire and black smoke roiling into the air. The colossal crash echoed down the canyon as the mangled rail cars and locomotive finally settled into the rocky stream at the bottom.

Oily black smoke billowing around him, Captain Ivanov sat on the tracks, his shoulders drooping and blood oozing down his chest. He coughed, shaking his head. Gathering his strength and will, Ivanov slowly turned to look west, but couldn't see anything through the smoke. He ejected the pistol's magazine and pushed in a fresh one, then slowly forced himself to his feet. Staring toward the spot where he'd last seen McClure, he staggered forward across the bridge. The American was gone.

CHAPTER 30

Elmendorf Air Force Base, Anchorage, Alaska

They all chose the meatloaf and mashed potatoes with gravy. Having perused the cafeteria selections and paid the cashier at the Iditarod Dining Hall, Novak, Mahler and Dantzig set their trays down together on a table near the far wall. The two tactical teams, twenty men, sat together at two adjacent tables. They nodded to Novak.

"This Air Force chow ain't too shabby, is it, sir?" Officer Tom Branson, one of the two team leaders exclaimed.

"Not at all, Tom. It looks pretty damn good," Novak replied, smiling.

"That salad bar is as good as anything I've seen," Meira agreed, taking a bite of the meatloaf.

Everyone was hungry. They had skipped breakfast to sleep in, although no one really got much sleep. An exhausting flight from Dulles into Elmendorf and anticipation of a gun fight against a Russian Spetsnaz special forces unit made sleep impossible. But halfway through his mashed potatoes, Novak's cell phone rang. Seeing the '703' area code, he stood abruptly and walked away from the table.

"Excuse me," Novak said to the group, lifting the cell phone to his ear and walking out of earshot.

"Pete, you there?" General Durwell boomed.

"Yes, Chuck, go ahead. We're just having lunch at Elmendorf. Nice place, this dining hall."

"Great. Look, I've got the information you needed. First off, the very last shipment, the one with the warheads, left Mountain Home Air Force Base in Idaho two days ago. It passed through the depot stop at Haines Junction in the Yukon roughly three hours ago."

"Thanks. And… anything else?" Novak asked.

"Yes. Pete, we've lost contact with the train. The tech people here think it's an antenna problem, but… ."

Novak interrupted. "Come on, Chuck. Antenna issue… my ass."

"I know, you're probably right. Should…?"

"Where would it be now, Chuck? How do I get to its last estimated location?" Novak blurted as he turned, glancing toward Raphael and Meira. Seeing his face, they slowly set their forks down on their plates.

"Well, assuming nothing happened in the Yukon Territory on the Canadian side, it should have crossed the White River and gone over the Canadian border by now. There's a bridge after that. I guess if I were planning to seize that train, it would be there, at that bridge. The locomotive has to slow down to cross the bridge."

"Right. Thanks, that at least gives us a vector. I'll take it from here. We're gonna roll pronto. Take care."

"Ah, Pete, assuming the worst, you know, a lost nuke, do you want me to declare a 'Broken Arrow'?"

"Absolutely not. We don't know anything yet. And the last thing I want to do is energize the whole damn defense department for nothing. I've got the stick, Chuck, and I'll handle it."

"Got it. Take care, Pete."

Novak slipped the phone in its belt holster and raised a hand, waving for attention. Raphael, Meira and the tactical squads looked up.

"We're rolling! Now!" He called out, spinning his right hand in a circle above his head.

The group rose from their seats en masse and moved to the exit doors, some of them jogging. Novak ran over and drew up alongside Officer Branson. "Call the air crews on your phone and tell them to get over to the flight line. We're dusting off in 20 minutes."

"Will do, sir!" Branson yanked out his phone.

"What about your gear, Tom, weapons, ammo?"

"All on board, sir. We're ready except for a few things still in our quarters."

"Super. We'll go to the VOQ first to pick those up. Divide your team up among the three choppers."

"Yes, sir!" They parted as they reached the doors.

The blue Air Force bus was waiting at the curb. As they climbed in, Novak shouted to the group, "We're going to make a quick stop at the VOQ! You have five minutes to grab what you need. Raphael, Meira, get your parkas and boots. Everybody, hustle!"

Novak turned to the driver. "VOQ, please, and wait for us. We're moving out pronto."

"Can do, sir." The bus lurched forward, took a couple of sharp turns and screeched to a halt in front of the VOQ. Nine minutes later, guards waved it through the gate to the flight line. It swerved left, crossed one runway and pulled up across from the three Black Hawks. Astonished, Novak saw that the pilots were already in their seats, the rotors turning slowly. The door gunners were adjusting the M60D machine guns on top of the heavy steel braces. The group piled out of the bus and split into groups as they ran to the three choppers. Novak, Raphael and Meira climbed onto the nearest one along with six tactical squad members, and strapped themselves in the jump seats.

The pilot swiveled in his seat to look back at Novak. Pete gave him a thumbs-up and the pilot relayed the launch order to the other two pilots. The powerful General Electric T700 engines roared, the rotor blades responding with a palpable increase in rotation, slicing through air. The sixty-foot long helicopters lifted off the tarmac in tandem and swung to the east, clearing the buildings and the base perimeter in just minutes. After climbing to two thousand feet, the Black Hawks leveled off. Novak unsnapped the harness and stepped into the cockpit.

"Way to go, Captain! All we're missing now is the 'Flight of the Valkyries'"

"Sir, that's a great scene! Robert Duvall is awesome in that movie! Especially his 'I love the smell of napalm in the morning'! I don't think there's a rotary pilot in the service who doesn't have that on a CD. You want me to play it over the loudspeakers?"

"No, no, really, I was just kidding. How long to the bridge, Captain?"

"We'll be pushing this to the max, sir, so I'm thinking about an hour and twenty minutes or so. Best we can do."

"Hey, that's great." He patted the captain on his shoulder and returned to his seat. "Raphael, Meira, you both okay?"

Both nodded back, smiling.

The thunderous roar of the engines and rotor blades made conversation nearly impossible. The three looked around the cabin, forward to the cockpit, aft and then to the doors. Meira had butterflies, never having been in a helicopter before, but the excitement of the flight far outweighed her nervousness. The Black Hawks edged up to 180 mph, zipping their way through the frosty Alaskan skies. At thirty minutes out, Tom Branson ordered the squads to check their weapons. Novak, Raphael and Meira turned to look as the six squad members jerked the slides of their Heckler and Koch submachine guns back and forth, ejecting and checking their mags. The three of them drew their pistols and did likewise.

"Mr. Novak, please come up here!" the captain shouted. "You need to see this!"

Novak unsnapped his harness and stepped to the cockpit, looking where the captain pointed through the windshield. Black smoke billowed up on the horizon slightly to the northeast.

"No! Damn!" Novak exclaimed, "Has to be the train."

The captain nodded. "We'll use a tactical approach, sir. Sweep our door guns over the target. They can chew up anything on the ground!" Novak stared in dismay, his hands gripping the seatbacks in front of him.

"Gunners, lock and load! This is a tactical approach. Flights break on my count... 3, 2, 1, break off!" The UH-60s on either side of them banked sideways, spreading to approach the bridge from different directions. Each Black Hawk swept low, making two passes over the area. The burning, smoking wreckage of the train at the bottom shocked them into silence.

"Sir, there appear to be several bodies on the track to the east," one of the pilots crackled on the radio. "I'll take a real low pass... they're not moving. What do you want to do?"

While their Black Hawk hovered away from the smoke and just off the east end of the bridge, Novak looked down through the door and saw the blood-spattered bodies of Mikulich, Sorokin and Zaretsky sprawled across the rails along the track. *What the...? Who the hell whacked all these Russians?* The thought hit him so suddenly he almost reeled. *McClure? Could it be?*

"Captain, believe it or not, those are Russian special forces, Spetsnaz. There may be more of them in the woods, so only one of choppers should set down and take a look. The other should be up on orbit for cover."

"Right."

"And it's not for sure, but it is possible that somebody I

know may be out here in the forest somewhere. He might be the one who put the kibosh on these Russians. He's the only one I can think of who could do this kind of damage by himself."

"Okay, I'll tell the crews we may have a 'friendly' out here. What do you want to do, Mr. Novak?"

"Keep one UH-60 up, put one on the ground to check things out. Ah, and tell the crew to be careful and stay away from the edge of the canyon. Put down to the east, there may be some radioactive contamination. Contact Elmendorf Air Force Base and request support… we need a nuclear response team out here."

"What, sir? Nuclear…?"

"Just inform your pilots, Captain. As for us, let's take this bird south toward the Bay of Alaska as quickly as we can. I've got a wild hunch."

"Got it, sir. Take your seat, please." The captain raised his mike to his lips. "Gunners, man your guns!" he snapped. The two gunners jumped from their seats, grabbing the overhead straps to steady themselves in the open doors.

Novak returned to his seat while the captain relayed instructions to the other two pilots. He waved off and abruptly pulled the Black Hawk into a 90-degree banking turn to the south, firewalling the engine. The UH-60 shot over the tops of the forest heading toward the sea, its gunners standing in their doors.

"Who did you say you know that may be down there?" Raphael asked.

"Dave McClure. Dave was with the Agency when he was younger than either of you, but switched over to the FBI when he got married. He lost his wife and child a year ago. It's a long story."

Raphael and Meira looked at each other, their eyes questioning.

CHAPTER 31

Alaska

He was dead tired, hurting all over. Searing pain clenched his shoulder and leg. McClure slumped on a box in the Russian camp, a scowl on his face and his chest heaving for air. It was a miracle he'd managed to trek all the way back here, carrying at least two bullets, a knife wound, and lightheaded from all the blood he'd lost. And hell, now it seemed as though it was all for naught. The train crashed, the nuke was gone, the camp was all packed up. He couldn't find any damn medical supplies, didn't even know where to start looking.

The somewhat good news was that the two bullet wounds in his shoulder and thigh appeared to have clotted, so the blood was only oozing now. The bad news was that the 9mm bullets were still in him. That definitely wasn't good. Plus, his makeshift sutures in the knife wound had ripped open, whether from exertion on the hike or in the frenzied gun fight at the bridge, he didn't know. The flesh had torn open again and it felt like the knife was still in him. It stung like a bitch. So much blood was smeared all over him that he wondered if it was even possible that it was all his. He thought he maybe had five of the original eight pints left. He shook his head groggily at the thought.

McClure looked at his watch… almost two in the afternoon. He unstrapped the snowshoes, took a long swig of water, and forced himself to his feet, weaving a little. He only had one round left in the rifle, but he edged the sling gingerly over his right shoulder anyway. You never know. Slowly and deliberately he picked his way down the ravine's incline through the empty camp and started down the trail to the shoreline.

It was cold, but not frigid. Off and on, the trees along the ravine blurred in his vision as he stumbled down the path. He paused to scoop up some snow, wiping it over his face, which helped a bit. Before long he saw the rocky edge of the shoreline and the water just beyond. He stepped cautiously out from the wooded trail, watching his footing on slick rocks near the shore.

McClure eased himself down to the snow, wondering what to do next as he watched the icebergs bobbing gently in the bay. Okay, an idea. He pushed unsteadily up to his knees and managed to stand. Some dry brush and branches stacked at the base of the biggest spruce he could find should do it. Ten minutes of scrounging yielded enough kindling, he hoped. He scraped a bunch of shavings off the magnesium block, then tossed a lit storm-proof match into the pile and backed up to watch. The fire caught and quickly blazed up the trunk of the spruce, smoking heavily. Maybe somebody would see it, that is, if anyone else was crazy enough to be this far out into the wilderness. He shrugged his shoulders at that thought. As his father often said, *'Nothing ventured, nothing gained.'*

McClure returned to his spot in the snow and lay down, looking out over the bay. Lots of ice floating on the gray surface today. He fought the urge to close his eyes and drift off… no pain, no worries, no more snow men… No. Shoving up to his elbows, then to sitting, took effort, but he managed it. He drank the last of his water and stuffed more snow in the bottle, then slapped his cheeks to dispel his lethargy.

What was that? His ears perked. He could have sworn he'd heard the high-pitched buzz from the Russian snow bikes off in the distance. That wasn't possible... or was it? And after that, he craned his neck, trying to localize what he thought was the distant whirr of a helicopter's rotor. *C'mon Dave, wishful thinking.* To be safe, he reached under his parka for the .44 magnum revolver, flipped the cylinder open and spun it... fully loaded. He lay back down, the pistol in his right hand on his chest, and struggled to take a deep breath.

"That won't do you any good. You'll be dead before you move a finger." The English was coarse with a strong Russian accent. McClure opened his eyes to see a snow man looming over him, a pistol pointed at Dave's head.

"*Privyet.*" "Hello." McClure pushed himself up again. "So, you speak English."

"Yes, and you speak Russian. And now... well, there can only be one victor in a battle, mister, ah...?"

"McClure. Dave McClure. And you are?"

"I am Captain Alexi Ivanov."

"Pleased to meet you, you Russian prick."

"And I am also pleased to meet an American asshole like you. I see you have a couple bullet holes in you. At least one of them is probably from my Grach MP-443." Ivanov smirked, then coughed, blood trickling from his lips.

"All expletives aside, Ivanov, I know what you shoot. But your own officer shot you. Looks like he punctured your lung."

"Yes, a lung shot. I admit, it's bad. My left lung is filling with blood. Major Krasnoff is not one of us. He's a worthless GRU bastard, but he won't live either. You shot him in the back with your, ah, Weatherby?"

"Yes, and you know your rifles. You and I are both going to die here, Ivanov. Neither of us will survive the arctic night."

"Perhaps, but you first. You killed all my men, McClure."

"And you bastards killed innocent men. That in your Spetsnaz code?"

"No, of course not." Ivanov frowned. "That's why I tried to stop that bastard Krasnoff at the train. None of that matters now, it's done. Enough of this. I admit you are a most worthy adversary, McClure. But as I said, there can only be one victor in battle. So, good-bye, asshole." Ivanov chuckled weakly and raised his arm forward, the pistol not three feet from Dave's forehead.

"Drop the gun! Drop it now!" Raphael Mahler shouted, stepping out from the trail.

Ivanov spun around, instinctively dropping to a crouch and sweeping the pistol's barrel toward Raphael. Two shots, both rounds went high and wide, thudding into a birch trunk. With a two-handed grip on his pistol, Raphael fired four times, the bullets striking Ivanov in the neck, temple and right eye. Blood, brains and other gore splattered as Ivanov was hurled backward and fell, his head cracking on an ice-covered rock.

McClure's eyes widened in surprise as he watched Ivanov fall, then slowly eased his head back down on the snow. Meira ran forward and fell to her knees next to him. Novak, his pistol drawn and six tactical squad members with submachine guns at their hips, swarmed out of the woods. His pistol still armed and ready, Raphael walked to Ivanov, looking down at the man's shattered skull. Meira lifted McClure's head into her lap and softly pushed back his hair.

"David," Meira whispered. She bent over and kissed his forehead. She looked into his eyes and kissed him again, stroking his hair while Raphael stood looking on. McClure smiled at her, his eyes gradually closing as he slipped into unconsciousness.

The Black Hawk soared over the trees and settled into the shallow waters forty yards to the west, the wash of its blades churning the water. Two crew members rushed forward with a stretcher between them. Edging Meira out of the way, they lifted McClure onto the pad with practiced care and strapped his chest and legs securely. As the men carried him to the chopper, Meira pushed herself to her feet and looked at Raphael who was staring at her with a strange look on his face.

"I'll be damned. It was Dave McClure after all," Novak grinned as the stretcher was carried past.

"Mr. Novak!" The med tech had to shout over the deafening thunder of the Black Hawk. "Sir, this man is badly wounded, appears to have lost a lot of blood. We're gonna get an IV started right away, but we really need to get airborne and get him to a hospital!"

Novak, Meira, Raphael and the team of six climbed on the Black Hawk, leaving Ivanov's body lying on the shore for later recovery by another crew. The UH-60 lifted sluggishly out of the shallow water and rose into the air, turning east. Pushing the engine full throttle, the pilot climbed to altitude. Aft, the med tech and another crewman huddled over the unconscious McClure, IV bags swinging overhead.

Novak sat next to Meira, across from Raphael. As the Black Hawk settled into its course, he shouted, "That was some damn good shooting, Raphael. Damn good. You saved Dave McClure's life. So, thank you."

"Luck, sir."

"Luck, eh? Well, Raphael, I've often thought about it and have come to the conclusion that when it comes down to it, I'd much rather have a lucky guy at my back than one who's just a good shot!" They both laughed, but Raphael's face quickly turned solemn.

"You have quite a mess to clean up, don't you, sir?"

"Indeed we do. I think we have a good chance of keeping it under wraps though. The Russians will have to offer an awful lot of concessions to keep this out of the international media. But I guess I'll worry about all that when I get back into my chair at Langley." Novak leaned back in his seat, staring somewhere over Raphael's head, then closed his eyes.

Raphael leaned forward across the aisle. "So, what was that all about?" he asked Meira in a moderate tone. Novak cracked one eye open.

"What's what about?"

"Why'd you kiss him twice? You kissed him once, okay. But why twice?"

"What?"

"Why two kisses? Gees, I thought you were gonna start making out with the guy."

Meira rolled her eyes and shook her head. "Okay, Rafi, because he's cute. He has gorgeous blue eyes, his name is David, and he's one helluva warrior, that's why. So, I kissed him twice. Rafi, the guy was almost dead. And hell, I can't wait forever."

Novak's eyebrows raised. He looked at Meira, then Raphael, trying not to laugh.

"You can't wait forever… for what?"

"Ah, we have a comedian. I'm not in any rush, Rafi, but I can't wait forever either." she sighed.

Novak couldn't help it. He began to grin, both his eyes open now and moving between the two.

Raphael frowned, "Meira… never mind"

"What? I can't hear you, what did you say?" Meira cupped her ears. Novak was now grinning ear to ear.

"Nothing. I'm sorry. I was just being stupid." Raphael shouted.

"Raphael Mahler… Rafi, you have nothing to worry about with me." Meira unsnapped her belt, stepped across the aisle, and kissed Raphael full on his lips. She moved back, easing herself down and smiling. "Nothing."

Rafi blushed and whispered, "Thank you, Meira Dantzig."

Novak leaned over to her. "Meira, you two keep this up and I'm afraid that you will make a wonderful couple. I can see it.

Oh, and anytime you think you might want a new job, give me a ring! You're good!"

Novak laughed again. Meira turned to him sporting a broad grin.

CHAPTER 32

JBER Hospital, Elmendorf AFB, Alaska

Pale light from the nurse's station illuminated the legion of tubes dripping fluid into McClure's arms. At two o'clock in the morning, the voices outside the room had trickled down to stillness. Pete, Raphael and Meira sat on chairs in the shadows of the room, staring at McClure's face. It had been six hours since Dave was wheeled out of the ICU and into this room. An oxygen mask covered his nose and mouth, the gentle hiss of the gas the only sound in the room.

Two shadows on the floor preceded the surgeon, still in his scrubs, and the ward's head nurse through the doorway. The nurse checked the instrument panels positioned on a rack above McClure as the physician walked over and stood in front of Novak.

"Welcome to the 673rd Medical Group. How's he doing?" the doctor asked in a low voice.

"He seems to be resting well. He hasn't flinched," Novak replied.

"After-effects of the anesthesia. He could wake up any minute, though. We had him awake briefly in the ICU before I felt he was sufficiently stabilized to bring him into the ward."

"At least one if not all of us will stay with him. It's late, doctor. Busy night?"

"Unusually so. All trauma injuries, nothing major, except of course Mr. McClure here."

"Can you tell me how you think he'll recover?"

"Sure. The biggest issue was the significant loss of blood. That affects the well-being of the whole physiologic and neurologic system. From the look of things though, my guess is that he'll be fine. We took two nine-millimeter bullets out of him, one in a shoulder and one in a thigh. Minimal damage there… he's lucky. The most serious injury was a three-inch-deep knife wound in his shoulder, so apparently he was involved in a hand-to-hand fight. It looked like he did a fair job of stitching himself up, but it was just normal thread and the sutures came apart later. Bottom-line, I would say Mr. McClure's a lucky man, but also tough physically with a strong will. We need to continue to hydrate him and put some more weight on him, but I think he's going to be fine."

"Thanks, doctor."

"You're welcome. I'll be checking in occasionally. Get some sleep yourselves, you all look beat."

Novak turned to Raphael and Meira as the physician and nurse left the room. "You two want to get some sleep? I'll be here."

"No thanks. The question is too important. We have to know," Raphael answered without hesitation.

"Okay."

The fatigue and dim light in the room soothed, but all three fought against dozing off. At three-thirty in the morning, McClure began to stir, rustling the sheets. Eventually, his eyes blinked open. Pete nudged Raphael and Meira, got up and leaned over Dave's bed.

"Dave, can you hear me? Dave?"

McClure looked up, squinting. "Yes… Pete, is that you?"

At that moment, a nurse walked briskly into the room. She

glanced at the overhead monitors, then poured some water into a glass on the tray.

"Mr. McClure, David, can you hear me, see me?" The nurse asked.

"Yes. Where am I?"

"You're in the hospital on Elmendorf Air Force Base outside Anchorage. Do you know where that is?"

"Yes."

"Great. I'm going to help you raise your head for a sip of water. Okay?"

"Sure." Her hand under his neck, McClure lifted his head enough to drink. She handed him two pills and he took another swallow to down them.

"Good. The sooner we can get you taking things orally, the sooner we can get these tubes out of you. Let me know if you need to get up. Here's the buzzer, just press the button to call."

"Okay." McClure drowsed. The nurse smiled at Novak and left the room. Pete drew back alongside the bed.

"Dave, Pete again. Feel up to a brief chat?" Raphael and Meira stood and joined him bedside.

"Go ahead, Pete."

"Dave, our teams can't get down to the train. It's way too hot, radioactivity. It will take some special equipment we'll have to fly in. We did find three bodies, apparently Russians, on the tracks near the bridge. Our team leader said they look like they'd been shot with a pretty heavy rifle round. That yours?"

"Yes. The snow men. I was in a firing position on the other side of the canyon. I used a Weatherby .300 magnum."

"Snow men?"

"Yeah. Well, they sort of looked like snow men to me when I first laid eyes on them, big bulky white parkas, white pants and dark goggles. I guess it sort of stuck with me."

"Okay, wow. Snow men. Dave, in searching the area, they

found two more bodies, same white parkas, in shallow graves near what looks like an encampment. You do that too.?"

"Well, directly and indirectly, but yeah, I did. Am I in trouble?"

"No, no, not at all. It's just…"

"There's one more near my own camp, southwest of the bridge, near my tent. It's all shot up though. Anyway, I buried him in a shallow grave too. Shallow is about all you can do in this tundra, and the best I could do in my condition. That one came hunting for me… at night. Tough bastard, ended up in a knife fight. He nailed me good in the shoulder."

"So that's where the knife wound came from?"

"Yeah. I stitched it but it didn't stay together."

"Gees, you took out six Russian Spetsnaz, decimated their team, Dave. Damn! Hey, remind me never to piss you off, okay?" They all laughed, even McClure chuckling weakly. "Ah, how many total?"

"Eight total. Pete, who shot Ivanov?"

"Ivanov?"

"At the shore… where you found me."

"Oh. Dave, this is Raphael and Meira, both Mossad. Raphael was the first to come down the trail onto the shore. He took the shot." Pete's hand landed heavily on Raphael's shoulder.

"I'd like to shake your hand, Raphael, but I've got too many tubes in my arms. You saved my life. Thank you. You obliterated the sonofabitch, must have hit him three, four times in the head. Anyway, thanks."

"You're welcome, Dave. My pleasure."

McClure turned his head slightly toward Novak. "Pete, I'm a little lightheaded."

"Okay. As soon as you're stable, they're going to move you to a hospital in Colorado Springs. And just so you know, we're picking up the tab on this. Hell, we still owe you. Just one more question, all right?"

"Sure."

"Did you see this guy among the Russians?" Pete held an 8x10-inch photo in front of McClure. "He's Major Vassily Krasnoff, a GRU officer, military intelligence, but we think assigned to lead this Spetsnaz team. Did you see him?"

"Yes. He was the eighth man. He was escaping from the bridge, heading east down the tracks on a snow bike, towing a sled. I nailed him. Looked like the round went into his lower back. I don't see how he survived it or even stayed conscious, but he lurched forward from the impact and then came back up and veered the bike into the woods. Out of my line of sight."

"Towing a sled?"

"Pete, I'm sorry. I think they took a nuke off the train before they crashed it, and put the warhead on the sled. I'm sorry, I tried."

Novak stood straight, his face instantly grave. He turned to Raphael and Meira, who slowly shook their heads but said nothing.

"Dave, you did far more than we could expect of anyone. What you did out there was... simply unbelievable, to say the least. Seriously. Please, excuse me for a couple minutes." Novak went out into the hallway, stepped near a window and pulled out his cell phone.

"Oh, no." Meira looked at Raphael, then took Dave's hand in hers.

Novak dialed long distance. It was ten-thirty in the morning in Washington, DC.

"Good morning, Joel Butler here."

"Joel, Pete Novak."

"Good morning, sir. Good to hear your voice. Are you okay? How did things go?"

"Joel, I'm fine. I'm calling because I've got an urgent task for you, very urgent."

"Yes, sir?"

"I need you to call Ted Rodgers over at Energy, DOE. He's in the headquarters on Independence Avenue. Ted runs Special Programs. Cathy has the number, just get it from her."

"Yes, sir."

"Joel, tell Ted Rodgers that this is extremely urgent. I need at least two, three if possible, of his NEST teams out here in Anchorage ASAP. They need to sweep the area northeast of Valdez in the Wrangell-St. Elias wilderness and east to the Canadian border. Critical."

"Sir, you said NESS?"

"No, N-E-S-T. Nuclear Emergency Search Team, NEST. It's very possible we have a nuclear warhead out there somewhere. If it's still there, that is. Talk to Ted Rodgers, nobody else. Take care and text me the status."

"Yes, sir. I'll get the number from Cathy and call right away. This is pretty bad, huh, Mr. Novak?"

"It could be very bad, Joel. Take care now." Novak signed off and walked back into McClure's room. Dave had dozed off again. Raphael and Meira sat against the wall, Raphael twisting his hands. Meira's eyes were wet.

"What now, Mr. Novak? Anything we can do?" Raphael asked.

"At least two Nuclear Emergency Search Teams will be flying out here later today. If the warhead is still out there in the wilderness, they can find it."

"And, if they can't? If it's not still out there?"

Novak looked at them both, concern wrinkling his forehead, and slumped into the chair next to Raphael.

"I don't have an answer for that yet."

CHAPTER 33

Langley Center, McLean, VA

The drifting fog and drizzle blanketing the Potomac Valley weren't enough to keep the Georgetown University crew team from plying the waters. The brisk shouts of the coxswain could be heard on both sides of the river as the long, narrow boat of rowers passed under Francis Scott Key Bridge, although it was too early in the morning to draw many passersby to the bridge railings. The usual deluge of traffic into Washington continued to cross the bridge as the boat disappeared into the mist. Miles to the north where the Potomac dwindled to a rock-strewn stream by the entrance to Langley Center, Pete Novak sat with gloomy face, watching the DCI walk back from the window.

"We seem to be breeding a long line of window watchers here, Jack," Novak said, taking a sip of coffee.

"What's that?"

"Well, Cathy often chides me for doing exactly what you're doing, walking back and forth to the window, the brain seemingly in lockstep."

"Well then, obviously great men think and act alike, Pete," Jack Barrett chuckled as he plopped into his leather executive chair and grabbed his coffee. "The Israelis got off okay?"

"Yes, we parted company at Dulles. They decided to take

advantage of a connecting flight to Tel Aviv. There was nothing more they could do here."

"True. I liked both of them. Great young people, extremely capable. When they marry, it will be interesting to see when one or both of them leave Mossad to start raising a family." Jack mused.

"It will indeed."

"And Dave McClure, you said he's been moved to Penrose Hospital in Colorado Springs? He was well enough for that? Gees. The man took two bullets and a knife wound. "

"Yeah, Dave's a resilient guy. He was recovering quickly, so the doctor saw no reason to keep him at Elmendorf. I'm actually more concerned about his psyche than about his physical healing. Then again, he seemed a different guy at the hospital, more optimistic. Could be that out in that icebox of a wilderness by himself, Dave found a way to deal with it all."

"You think he'll take you up on your offer to come back? The way you talk about him, Pete, he sounds like he'd be one hell of an asset."

"I'm hoping." Pete paused and looked Barrett directly in the eyes. "Jack, we're hopping all around it with this small talk. You know I'm worried all to hell about this. I need to hear your take. The NEST team followed a radioactive trace down to the shore, miles east of where we were. That means the nuke got transferred to a vessel. The Iranians will have the nuke, Jack, and hell, you know they'll use it. You know that. Can't we take some action?"

"On the high seas? Pete, this isn't the 18th century and we're not the Barbary pirates. We can't go boarding every ship crossing international waters. It may have been moved to air transport at some point anyway."

"So, that's it. There's nothing we can do. It's a fait accompli."

"I didn't say that, Pete. We still have options. It just depends on the willingness of the people at the top to exercise them."

Barrett pushed his chair closer to the desk and gulped down the last of his coffee.

"What? I don't understand. What options do we have now?"

"Pete, you're not... oh, hell." Barrett hit the intercom switch. "Marianne?"

"Yes, sir?"

"Please print out a standard SCI Briefing Acknowledgement and Nondisclosure form with Pete's name and data on it. The program is 'Meridian'."

"Yes, sir."

"Meridian?" Pete repeated.

"I can't tell you everything about it, but..." Marianne walked in and handed the nondisclosure form to Barrett.

"Thanks, Marianne."

"You're welcome, sir." As she walked away, he pushed it in front of Novak.

"Go ahead and sign it, Pete. In a nutshell, and that's all I can give you, a nutshell, Reagan's Star Wars initiative didn't die a lonely death back in the '80s. Sure, the Soviet Union collapsed and the strategic threat seemed mitigated, but the program was kept alive and went into the black R&D world, that's all. Deep black. Now, thirty years later, it's complete, fully deployed and fully operational. Think back, you should remember this. As I recall, you had some interesting involvement with this back at the beginning."

"I do remember. I just thought the whole thing was dead."

"Nope. Alive and well."

"So, you're saying we have... ah..."

"We have constellations of maneuverable satellites in varied orbits. A mix of hydrogen fluoride lasers, particle beam, kinetic energy weapons... they call those platforms 'smart rocks', 'brilliant pebbles'. Space-based ballistic missile defense. Nobody knows we have it, nobody. And the president has mandated that no one reveal that we do have it. Period. Furthermore, it takes Presidential authority to exercise it."

"Really? Gees. So, if Iran launches a Safir II missile with a nuclear warhead, we could kill it?"

"The operative word is 'could'. Oh, and I was going to brief you on this today anyway. The president has ordered four carrier battle groups to the Gulf. They'll be steaming in there by next Saturday. That's four carrier battle groups… the USS *George Washington,* the *Harry Truman,* the *Ronald Reagan…* she's gorgeous, gigantic, and the *George H. W. Bush,* along with their escort warships. It's the greatest amalgamation of naval sea power since the Third Fleet in WWII. They can send the nation of Iran back to the Stone Age."

"Back to the Stone Age, heh? Well, that'll only set 'em back two weeks. Jack, you know and I know that those carrier groups aren't much of a deterrent if the Iranians think they can destroy Israel in the process."

"Israel has very effective ground-based BMD. We gave it to them. That should be enough to handle most anything."

"So? Jack, if the warheads are salvage fused so they detonate on destruction, there would still be humongous nuclear air bursts. The effect on Israel will be the same: *Adios Muchachos.*"

Barrett nodded. "I agree with you on that, Pete. But like I said, the decision authority rests with the president. And so far, he hasn't budged on that subject."

"Okay, thanks, Jack. I asked for your take and I got it. I have to say though, it's a sad state of affairs. I hate to see everything end up like this."

"I agree. Look, take the rest of the day off and get the hell out of here. There's nothing going on and I've got your back. Go take Sarah out to lunch."

"I just may do that, Jack. Thanks." Novak rose, pushed up from his chair, gave a half-hearted wave, and left Barrett's office the same way he'd entered, in a state of gloom.

CHAPTER 34

Washington, DC
December

The first snow of the season fell gently on the nation's capital in soft, downy flakes. The first week in December. Kids hoped there would be enough snow this year for Santa to get through with his sled full of toys, and parents throughout northern Virginia, Maryland and the District assured them there would be. Santa never failed to deliver the goods. The National Christmas Tree had been lit a week ago on the Ellipse just outside the White House gates, with the president and his family officiating.

The holiday shopping season began to bloom in all its economic stimulus glory. Jammed with shoppers, malls around the metro area were decked out in all the usual holiday trappings of holly and ivy and special lighting. In the Jewish community, the faithful were preparing for the eight joyful days of Hanukkah. The nation eagerly turned its attention toward the perennial thoughts of good times, family, tradition and the holiday season.

Just past midnight on December 7th, the phone rang 1, 491 miles away from Langley Center at a controller station at the NORAD Command Center in Colorado Springs. The controller, Air Force Captain Herb Walgren, instantly linked the call to the Space Defense Operations Center (SPADOC) and the Missile

Defense Agency, activating the national Integrated Tactical Warning and Attack Assessment (ITW/AA) system.

"SPADOC, I have notification of a launch at latitude 35.234 North, and longitude 53.921 East... that's Semnan, Iran. Do you copy?" The controller spoke in monotone.

"That's affirmative, NORAD, this is SPADOC. The launch is an announced commercial satellite launch. We're tracking. SPADOC out."

"Do you have telemetry data available yet?" The NORAD controller asked.

"Checking. Stand by," the young voice at SPADOC replied.

"SPADOC, I have indications that this launch is not transferring to a commercial orbit. Can you confirm?"

"Roger, that's affirmative, NORAD. We show that too. Telemetry indicates this launch may be headed to the eastern Mediterranean basin. Repeat, launch may be hostile. Three minutes, twenty-four seconds to impact. Continuing to gather telemetry."

"Will hold, waiting for confirmation, SPADOC." Beads of sweat blossomed on Captain Walgren's brow.

"Launch telemetry is confirmed, NORAD. Characterization is Safir Class II rocket headed for eastern Mediterranean. The target is Israel. Confirm, missile is Safir Class II rocket. Assume missile is hostile. Target is Israel. Two minutes, fifty-three seconds to impact. Do you read?"

"Affirmative, SPADOC, sensors confirm telemetry," Captain Walgren acknowledged.

"NORAD, recommend immediate CINCNORAD and CINCSTRATCOM notification. SPADOC out."

"Roger. Executing." The captain hit several buttons in quick succession on his console. "General Whitcomb, sir, do you read me? This is Captain Walgren, duty officer."

"Yes, captain. Go ahead." The general had grabbed the gray

phone on the night stand, sat up in bed and turned on the lamp, blinking in its glare. He glanced at his wife, who just rolled over and pulled the covers up. They had both hit the sack early after returning from a holiday party. The two-story Broadmoor Bluffs house, in the shadow of the Cheyenne Mountain Complex just southwest of Colorado Springs, received these late-night calls all too often for her to be concerned.

"Sir, we have a confirmed launch at Semnan, Iran. SPADOC confirms, indicates launch was announced commercial, but sensors confirm missile is headed toward Israel. Assumed hostile intent. We have two minutes, twenty-five seconds to impact, sir."

"Got it, Captain. Damn. Captain, you are authorized to notify NMCC immediately. Repeat, notify NMCC. Back brief when possible, please."

"Yes, sir." Captain Walgren keyed the alert line for the National Military Command Center, J-3 Operations at the Pentagon in Arlington, Virginia.

"Receiving. Go ahead, NORAD... NMCC responding, Colonel Moretsky here." The voice was male, mature, unemotional.

"Sir, NORAD and SPADOC confirm hostile launch of Safir Class II rocket from Semnan, Iran, to Israel. Impact is estimated at two minutes, three seconds."

"Crap... ah, that's affirmative, notification received. NMCC will execute National Command Authority notification. NMCC out."

Another bedside phone, this one belonging to the Chairman of Joint Chiefs of Staff, rattled off the hook in his home at Fort McNair at the end of the peninsula between the Anacostia River and the Washington Channel.

"General Holmes, here." The Chairman answered.

"Sir, this is Colonel Moretsky, NMCC. This is official NCA notification, sir. Hostile missile launch from Semnan, Iran.

Sensor-gathered telemetry indicates target is Israel. We have a parallel MDA alert message that indicates three Meridian platforms are prepared to interdict. Repeat, Meridian system platforms are prepared to interdict. Request authority to execute, sir. One minute forty-two seconds to impact."

"Notification received, Colonel. Keep this line open. I will notify the president and reply."

"Will do, sir."

An aide tip-toed into the bedroom, gently shook the president awake and led him out into the hallway. He handed the President the phone. The president shivered slightly as his bare feet hit the cool floor, and did a little jig in place.

"This is the President."

"General Holmes here, Mr. President. Sir, this is NCA notification. Hostile launch confirmed from Semnan, Iran. Target is Israel. Meridian system is in place and can interdict. Request authority to execute, sir. One minute fourteen seconds to impact."

"General, Israel has ground-based ballistic missile defense." he paused. "We gave the system to them. I'm against compromising the Meridian sys…"

"Mr. President, this missile… it may be nuclear, sir, and if it's salvage fused, ground-based BMD won't help. Eight million lives at stake, sir. We have fifty-one seconds to impact, Mr. President."

"General, I said I'm not going to…" The Chairman sighed on the other end and the line suddenly clicked off and went silent.

"What? What happened?" the president shouted, then screamed into the phone, "General Holmes! General, you bastard, don't you dare hang up on me! General! General Holmes!" His cold feet dance became a furious stomp on the wooden floor. His aide turned white at the sudden display of rage.

"Colonel Moretsky, do you copy? This is General Holmes,

Chairman, OJCS."

"Yes, sir, I copy."

"Authority to exercise Meridian granted. Kill the sucker."

"Yes, sir."

"Oh, and Colonel?"

"Yes, sir?"

"I'm going off-line for a bit now. When the White House calls, please congratulate the president on my behalf for killing a hostile Iranian missile. I was getting tired of this fucking job anyway. Holmes out."

"Sir, what?"

The line was dead. Moretsky sprang into action. It took him roughly eight seconds to execute the order. Three watermelon-sized chunks of tungsten shot from their platforms and hurtled through space at 28,000 miles per hour. The first 'smart rock' struck the nose cone on the last stage of the Safir II missile just as it began opening to deploy the MIRV warheads. The nose cone shattered, breaking apart as the second and third kinetic-energy weapons struck, obliterating the launcher and the MIRVs. Three expanding orbs of brilliant light blazed in thermonuclear explosions across exospheric space.

Vice Admiral Tom Ramsey stood forward on the bridge, holding a cup of coffee and staring out the steel-framed windows of the USS Ronald Reagan as his flagship of the carrier strike group steamed through the Persian Gulf. The seas were easy and the air cool.

"What's our current position, Nick?" he asked.

"Sir, Bandar Mugam is 29.6 miles off the starboard bow." Lt. Commander Nick Getty replied.

The admiral glanced at his watch and smiled – 1:35 PM. Right on the timeline. "Good. Come left to the new heading of 280 degrees west-north-west. Transmit on NAVCOM to the group."

"Yes, sir. Come left 280 degrees west-north-west... transmitting."

In a sudden instant, the sky to the carrier's stern gleamed a stark bright white. An expanding rim of searing yellow, red, and white light flared across the heavens, the four steel-gray aircraft carriers and their escort ships glimmering in the sea with a surreal chrome reflection. The simultaneous thermonuclear flash of the three MIRV warheads detonating in space appeared as though the sun itself had exploded.

"What the hell?" the admiral gasped. "Sound general quarters!" he barked in a fluster.

Near instantaneously, the blaze of light began to recede as quickly as it had begun. The skies above reclaimed the blue hue and the sun shined above them once again.

"What the...?" Ramsey exclaimed. "What just happened? Ah, cancel that order, Nick. Notify NAVEUR, 6th Fleet, of what we've just observed. Ask them if they know what occurred."

"Yes, sir."

The admiral's eyes glued to the windows, the strike group continued west-north-west to its staging area. Remaining dusty debris in exospheric space immediately began to sink in a deteriorating orbit, to burn up in the atmosphere within hours.

People were mulling about on the beach in Tel Aviv, having seen the bright flash in the sky and discussing among themselves its possible origin. The sun was now shining, glittering on the waters of the Mediterranean. The phone rang off the hook at Mossad Center. A tired and disheveled Taavi Perutz looked over at General Ariel Goren, set his coffee mug down, and lifted the receiver.

"Sir, Peter Novak is on the line 1 for you." The receptionist said.

"Put him through."

"Yes, sir."

"Hi Pete, this is Taavi. How're things in Washington?"

"Taavi, I hope you're sitting down. If not, please sit down."

"Yes?" Taavi's brow furrowed instantly.

"I called to tell you that the nuclear warhead the Iranians had is dead, Taavi. We killed it thirty-six minutes ago in exospheric space. It's gone. Destroyed."

"Pete... my God, that's fantastic!" Perutz glanced at Ari. His eyes grew wet. "They launched it?" General Goren leaned forward in his chair with a sudden look of concern.

"They did indeed, Taavi. And they're gonna pay dearly for that."

"God be with you, Pete. I'm sorry... but..." His chest heaved and a muted sob broke from his lips.

"It's okay, Taavi. I understand. We'll have four, that's four, carrier battle groups in the Gulf within forty-eight hours. We haven't formed a naval force that big since World War II. There's going to be hell to pay. You might want to put your forces on alert. I'm sure notification is headed your way through official channels, but wanted to tell you myself. Shalom."

"Shalom, Pete." The line went dead.

Perutz came around his desk, pulled Ari up by the shoulders and wrapped his arms around him.

"The Americans destroyed the nuclear warhead, Ari. It was launched. It was on its way here, but it's gone. It's over."

In the parking lot outside the center, Raphael leaned against his car, looking out to the sea. Meira stood next to him, her head snuggled against his chest. His cell phone rang.

CHAPTER 35

Phuket, Thailand

Sweet scents of jasmine, gardenia, orchid and champak wafted on soft breezes flowing off the azure Andamen Sea. At two o'clock in the afternoon on December 28th, throngs of people strolled in and out of restaurants, bars and shops fronting the palm-lined intersection of Bangla and Thaweewong Roads in Patong Beach. A few sandal steps away, powdery white beaches hosted a legion of bikini-clad sunbathers reveling in the balmy eighty-one degrees. The Thai slogan '*Mai pen rai*', 'Don't worry, be happy', was alive and well in Phuket.

Even among the exquisite environs of Patong Beach, meandering tourists on the shady promenade gasped in wonder at the entrance to the Royal Orchid Resort. The five-star beach resort offered the epitome in sumptuous luxury, its white Roman columns reflecting the radiance of the afternoon sun and advertising the highest standard of lavish, indulgent hospitality found within.

On the back side of the resort, facing the sea, Kurt Hartmann relaxed, stretched out on a chaise lounge amid the manicured lawn, thumbing through the pages of the *Wall Street Journal,* a complimentary perk made available daily to guests at the Royal Orchid. No small achievement given Phuket's distance from New York City. He checked the status of a few stock investments he'd effected in the West's capitalist markets.

The young woman appeared from a rear door of the hotel and strolled casually down the slate walk that wound through richly flowered gardens lolling in the mottled shade of palms. She deftly balanced a single Mai-tai cocktail on her tray, the turquoise bikini and stiletto heels accentuating her firm calves and slender thighs. Joisette's brown hair bounced lightly off her shoulders as she approached him, bronze skin glistening in the sun.

"Sir, I'm sorry to interrupt you, but did you order a Mai-Tai? The bar mixed up some orders and we seem to have an extra," she spoke softly, apologetically.

Kurt looked up from his paper momentarily, caught off guard by the woman's stunning beauty. "Ah, no, I didn't but I'll take it off your hands. It's quite warm out here."

"Thank you. Here you go." She set it on the glass topped rattan table nudged next to his lounge, and turned back toward the hotel.

"Wait, please don't go. I'm Kurt Hartmann. You are?"

"Joisette Fracassi. I'm pleased to meet you."

"Well, you're obviously not Thai. And my, Joisette, may I say that you look totally delicious in that bikini. Where are you from?" He smiled blatantly, admiring the curves of her body.

She blushed visibly. "Paris."

"Really? Where exactly?"

"Neuilly-sur-Seine."

"Nice area."

"My father was a physician."

"Your father's a doctor, and you work here? Why?"

"I'm an artist and it's an exotic locale in Phuket. I work in oils and acrylics, and the colors here are remarkably luscious. Where're you from?"

"I'm from Wiesbaden. West of Frankfurt. My parents were

Russian emigrants. I know you can still hear that in my voice. But I lost my brother recently... to kidney failure and a coma. After they finally removed his life support, well, I needed to get away, so I live here now. Joisette, what would you say to dinner tonight?" He placed his left hand lightly on her calf.

"I'm sorry to hear about your brother. Well... are you a guest here?"

"Not really. I have a penthouse residence on the fourth floor. So, you wouldn't be violating any hotel protocol by having dinner with me. What do you think?" Kurt's hand slid gradually up to her thigh, gently caressing.

"You have a penthouse here?" She tightened her muscles, letting him feel the firmness of her legs.

"Yes, I do. So, interested?" He smiled, his fingertips probing under the taut fabric of her bikini.

"Well..."

Kurt gently squeezed her cheek.

"Ah..." she blushed. "Okay, yes, I think I would."

"Great. Then shall we say, seven o'clock, suite 407, Joisette?" His hand fully under the bikini now, he squeezed her tenderly again, and licked his lips.

"Yes, I'll be there," she smiled, gently pulling his hand out.

Kurt turned his head, watching the swing of her ass as she walked away, and grinning broadly. He would have a fantastic dinner with Joisette tonight, topped off with an enormously satisfying... dessert. He reached for the Mai-Tai and raised it to his lips as Joisette disappeared into the garden. He took another sip, the slight under taste masked by the 151-proof rum.

"Whoa. "What the hell?" Kurt shook his head as a bizarre dizziness swept over him, followed by sudden weakness in his hands and arms. The Mai-Tai slipped unexpectedly from his grasp, striking the table, shattering glass on the lawn. He grabbed the arms of the chaise lounge and forced himself to his feet. Gasping, he tried to breathe. But there was no air, and his head felt like it was going to explode. Without warning, he collapsed to his knees in a violent convulsion, his face coloring pinkish red as the cyanide prevented his body from using the oxygen in his blood. With a last glance to the sky, Leonide Krasnoff crumpled face down in the grass. Nearby guests stood abruptly and looked at him in alarm.

From behind a palm, Meira Dantzig watched the Russian fall to the lawn, stone-cold dead from the cyanide laced cocktail. She dropped the tray on a chair, tugged off the high heels, and tossed them carelessly into the bushes, slipping her feet into colorful flat leather sandals and pulling a feathery linen cover-up over her bikini. Meira took one last glance over the foliage at the expired Krasnoff. She shrugged her shoulders and strode away through the lush greenery of the Royal Orchid Resort.

CHAPTER 36

Paris, France

Lunchtime aromas drifting from the kitchen at the Café Tante Louise were nearly irresistible. Even people hurrying by stopped to sniff hungrily, their mouths watering. Ever since Leonide Krasnoff introduced the Café Procope to him, Bizhan Madani had grown quite fond of the restaurants along the Rue de l'Ancienne Comédie. And while the Café Tante Louise, across the street and three doors down from the Café Procope, didn't have a historic artifact on display like the Café Procope's writing table of Voltaire, the cuisine was just as good, perhaps even a little better.

This was Madani's third meeting there: just with some people from the embassy, which nonetheless, gave Madani another opportunity to visit his favorite western city, Paris. A table for four, encompassing lunch, wine and a smidgeon of intelligence chit-chat. He'd ordered the lobster-crab bisque to start off, delicately fortified with a touch of sherry, beyond scrumptious. His unrefined colleagues ordered the more bourgeois cream of potato soup.

Even in the City of Light, it seems, money talks. For the handsome sum of five hundred euros, a member of the wait staff had agreed to mix a strong laxative into Madani's bisque, adding an extra dash of sherry to preclude suspicion. After all, this was just a diplomatic gag. As Madani spooned up the last of the delightful bisque, he felt rumblings begin in his stomach. Ignoring the slight discomfort, he eagerly moved on to his beef burgundy over noodles. Again, as expected, the dish was wonderful.

As the waiter cleared the table, Madani felt he had to pass on coffee and dessert, squirming uncomfortably in his seat. Pressure built in his lower colon, causing awkwardly unpleasant distress. In hushed voices, the group discussed yet another Iranian emigrant in Great Britain who'd been persuaded to work with Vevak, and had recently received instructions to apply for a Farsi translation position with the British Secret Service. Madani was indeed interested, but could sit still no longer. The pressure in his bowels had become unbearable.

"*Pardonnez moi, mes amis. Un moment,*" he exclaimed, rising abruptly from his chair. His colleagues lifted eyebrows in surprise, but nodded their heads in response.

As Madani walked hurriedly to the men's room, his security escort automatically jumped from his post near the door.

"I need to go to the restroom right away, Farjad. Please."

"Sir, let me check it quickly." Farjad found the men's room empty. He rapidly examined the urinals and swung open the door to the toilet. Nothing. Satisfied, he walked back out.

"It's clear, sir. I will wait here and bar any further entry."

"Thank you." Madani rushed in with obvious urgency, ran to the toilet, and slid the bolt across the brass plate. Practically ripping off his pants, he plopped on the seat to relieve himself. A smile of release washed across his face, but the relief was short-lived. A Mossad tech team had entered the lavatory for a minor renovation just after he arrived at the restaurant. The toilet seat where he now sat was comprised of an unpleasant mix of semtex, thermite and white phosphorous.

Seated at the bar next to Dani, Raphael had allotted roughly thirty seconds for Madani to drop his drawers. Glancing up from his watch, Raphael nodded, and Dani pressed the detonator sending the signal to the electronic trigger in the seat. Although the carefully estimated chemical mix made more of a loud whooshing sound than an explosion, the bright light of the thermite and phosphorous incendiaries radiated through the cracks around the door.

Farjad rushed immediately to the door, but the searing heat was unendurable. He backed up, watching in horror as the wooden door was scorched black. The sprinkler system flashed on, showering the café patrons with a cold spray. Screams erupted as terrified customers tripped over each other, scrambling to flee the restaurant. The beautifully framed pieces of modern art in the lavatory were gone, the walls now adorned only by charred pieces of Bizhan Madani.

"Where to Rafi?" Dani asked a few minutes later, his hands on the steering wheel of the Audi.

"'Rafi'? You too?"

"Everyone seems to be calling you that now. I kind of like it."

Rafi shrugged and patted Dani's shoulder genially. "You know, Dani, we probably won't see Paris again for years. Let's stop somewhere on the way to the airport and have a nice, chilled glass of Chablis. My treat. We have more than enough time." Sirens wailing already just blocks away, the Audi swerved from the curb and sped down the avenue.

CHAPTER 37

Colorado Springs

McClure sat in his leather easy chair in the family room, staring out through the window toward the mountains and Pikes Peak, though his eyes were unfocused. The only movement was an occasional blink. The coffee on the end table had gone cold an hour ago.

He was two thousand miles away, watching the trail ahead as he trekked through the snow. He jigged to the right, then left as his snowshoes broke through the icy crust. Scenes of the deep Alaskan forests rolled across the screen of his mind: aspen, birch, black spruce towering over him amid glistening drifts of snow; caribou, bears, and snowy owls appearing and moving on; black capped chickadees and gray sparrows hopping around tree trunks, pecking at the tips of grasses. In another instant flash of memory, he jutted his head through the flap of the tent, looking up at the millions of stars sprinkled across a black night. A smile played across his lips at the memory of the dark arms of trees reaching upward, buttressing the sky.

The doorbell startled him back to reality at four o'clock in the afternoon. McClure walked across the room, a noticeable limp in his gait, to throw the deadbolt and unhook the chain. Gus trailed at his heels.

"Happy New Year!" Pete Novak stood on the porch, waving

one hand in celebration and hefting a bottle of Moet et Chandon champagne in the other. Bundled up in an attractive black cashmere overcoat and a white scarf, he shivered against the bitter winter wind.

"Mr. Novak, Pete, happy New Year to you too. Another meeting at Air Force Space Command?"

"Yeah, the meeting broke early. I'm not flying out until tomorrow, so… I bought a bottle of the good stuff and thought I'd pay you a visit. The champagne is straight from the caves at Epernay, France."

"Please, come in."

Pete stepped past Dave and went straight to the family room, McClure limping his way behind.

"Glasses?"

"In the kitchen, I'll get…"

"No, Dave. You go sit down, just tell me where they are."

"Left cabinet, top shelf."

"What's that you've got playing on the Bose?"

"George Winston. It's his '*December*' album, solo piano, a piece called '*Thanksgiving*.' It's… beautiful, restful."

"Indeed it is." Novak paused, watching McClure sprawled in the chair, once again staring out the window at the mountains to the west. Pete's face grew solemn, his brow furrowing. He hoped he wasn't missing anything. Returning to the family room with two glasses, he popped the cork and poured. He began to raise his glass, but hesitated when Gus pawed at his leg, tail wagging feverishly. Gus tilted his head, giving Pete a curious look and a yip.

"Who's this?"

"This is Gus." At the sound of his name, the pooch ran and jumped in McClure's lap. Dave rubbed him vigorously behind the long ears. "Beagle mix. I went over to the Humane Society one morning, on a whim really, you know, just to look around. And well, when Gus and I saw each other, it was love at first sight. Right, boy?" McClure found a sweet spot and Gus squirmed in ecstasy. He and Novak both laughed at the sight.

"Had him long?"

"Just a few days, actually. But the truth is, Pete, I love having him around, and he seems to like being here."

"Great idea, really. Well then, here's a toast to you, me and Gus! Happy New Year!" The two raised their glasses and took a sip. Gus leaped from McClure's lap and trotted away, disappearing around a corner.

"Gus is still exploring the place. It's a gas to watch him, to see his reaction when he finds something new." Dave chuckled.

"Good for him and good for you too. You saved a lonely pooch."

"That I did. Pete, how are things at the Agency? Especially the aftermath of the Alaska thing?"

"The Alaska thing… that's as good a way as any to describe what happened… actually an unbelievable series of events. The train, well, the crew and the security force were all dead. The forensic techs said they died in the crash, not from the intense radiation. Some degree of relief in that, I guess. And the Russians are falling over themselves to make amends. They're cleaning house big time, lots of firings and dismissals over there. Oh, and we killed the warhead."

"Really? How'd you manage that?"

"Can't tell you." Novak laughed. "I'd have to kill you."

McClure chuckled. "That's okay. What about… oh, Raphael and Meira?"

"You have a good memory, Dave. They're fine. By the way, Major Vassily Krasnoff died from your bullet. Our sources say it tore out a kidney and a large piece of his intestines. Word is, he went into a coma, then kicked off several days later. That means you whacked seven of an eight-man Spetsnaz team

– by yourself, in the middle of the Alaskan wilderness. That's a modern *David and Goliath* tale if I ever heard one, pun intended. Certainly, it's one for the record books. Oh, and Major Krasnoff's brother, Leonide, the Russian mafia guy who set the whole caper up, he's disappeared from the landscape in Phuket, Thailand. So did the Vevak intelligence contact, Bizhan Madani, about a week later in Paris."

"May I ask how that happened?"

"Mossad."

"Makes sense."

"Dave, how's everything else? You know..."

"I'm fine, really. You were right, I had to find a way to live with my past, and I have. Alaska did that for me. You know, Pete, I think part of me is still up there in the wilderness. It was such a profound experience."

"I can only imagine. Well, fantastic. I'm very glad to hear that, Dave. I think the guy who hiked into those woods on snowshoes is not the same guy who came out on a helicopter. You definitely seem different. Any nightmares? You know, all the violence, the killing?"

"No, not really. That's faded into the background somehow. Pete, you know, all the years of training, the constant practice, well, my years in the FBI steeled me to it, I guess. I am surprised a little bit about it myself, though. I mean, all of that, it's there, somewhere in some corner of my memory, but what I really remember most is the forest, the woods. The overwhelming beauty, peace, stillness... it's really something, Pete."

"I can only envision through your eyes and your telling of it, Dave. I'm glad those are the things you remember most. So... I have to ask you, would you like to come back? You obviously haven't lost your edge, far from it. We'd love to have you. I'd start you as a Section Chief, a lead, in Special Operations."

"Pete, let me ask you for a rain check on that. Give me, oh, six months. By this summer, I should have a handle on what I want to do with the rest of my life."

Novak stood and pulled Dave up, looking into his eyes with a smile.

"Fair enough. Well then, Dave, here's to us... you, me and Gus. To a very happy new year!" They raised their glasses in a toast to the future.

The End

Acknowledgements

Most especially, I would like to offer my sincere thanks to my editor, Ginny Ruths, Touchstone Publications, for her brilliant work in strengthening the narrative and plot flow. Also, heartfelt thanks to my friend and college classmate, Darryl Wimberley, a well-known author himself and winner of the Willie Morris Award for Southern Fiction, for his frequent snippets of advice regarding plot development. I offer sincere thanks to my reviewers: Joe Dalton, John Fraser, Edward MacGuire and Tom Owen for their critical reading of the original manuscript and their many editorial suggestions. I would like to add very special thanks here to my wife, Trudi, for her support in my commitment to the project and who also offered an abundance of suggestions for my writing style. Thanks as well to my friend, Sergey Pancheshnyy, for his kind assistance in my understanding of the Russian language. Tremendous thanks goes to Leigh Coates and Wyatt Ratterree, both pilots with Vertical Solutions in Valdez, Alaska for an incredible helicopter flying adventure over rugged wilderness and landing on glaciers which in fact planted the seed for this story. Finally, thanks to the wonderful people of Alaska and their governors, past and present, who together have worked so diligently to protect and maintain the pristine wilderness that makes the state of Alaska such a magnificent place to visit.

CPSIA information can be obtained at www.ICGtesting.com
Printed in the USA
LVOW05*0536210715

447006LV00012B/63/P

9 781478 744573